Behavior and Arteriosclerosis

Behavior and Arteriosclerosis

Edited by

J. Alan Herd

The Methodist Hospital
Houston, Texas

and

Stephen M. Weiss

National Heart, Lung, and Blood Institute
Bethesda, Maryland

Plenum Press · New York and London

Library of Congress Cataloging in Publication Data

Main entry under title:

Behavior and arteriosclerosis.

Includes bibliographical references and index.
1. Arteriosclerosis—Psychological aspects. I. Herd, J. Alan. II. Weiss, Stephen M.
[DNLM: 1. Arteriosclerosis. 2. Behavior. WG 550 B419]
RC692.B43 1983   616.1'36'0019   83-10932
ISBN 0-306-41281-0

©1983 Plenum Press, New York
A Division of Plenum Publishing Corporation
233 Spring Street, New York, N.Y. 10013

Printed in the United States of America

Contributors

FLOYD E. BLOOM, M.D.   Professor, The Salk Institute, P.O. Box 85800, San Diego, California 92138

JOSEPH V. BRADY, PH.D.   Professor of Behavioral Biology, Division of Behavioral Biology, The Johns Hopkins University, Baltimore, Maryland 21205

ROBERT S. ELIOT, M.D.   Professor and Chairman, Department of Preventive and Stress Medicine, University of Nebraska Medical Center, Omaha, Nebraska 68105

RICHARD I. EVANS, PH.D.   Professor of Psychology, Department of Psychology, University of Houston, Cullen Boulevard, Houston, Texas 77004

JOHN W. FARQUHAR, M.D.   Director, Stanford Heart Disease Prevention Program, Stanford University Medical School, Stanford, California 94305

DAVID C. GLASS, PH.D.   Professor of Psychology, Graduate School and University Center, City University of New York, 33 West 42 Street, New York, New York 10036

L. HOWARD HARTLEY, M.D.   Associate Professor of Medicine, Cardiovascular Division, Brigham & Women's Hospital, 75 Francis St., Boston, Massachusetts 02115

J. ALAN HERD, M.D.   Medical Director, Sid W. Richardson Institute for Preventive Medicine, The Methodist Hospital, 6565 Fannin, MS S400, Houston, Texas 77030

JOSEPH D. MATARAZZO, PH.D.   Professor of Medical Psychology, Health Sciences Center, University of Oregon, Portland, Oregon 97201

PEKKA PUSKA, M.D.   National Public Health Institute, Mannerheimintie 166, 00280 Helsinki 28, Finland

TERESA SEEMAN   Doctoral Candidate, Epidemiology Program, Department of Biomedical and Environmental Health Sciences, School of Public Health, University of California, Berkeley, California 94720

S. LEONARD SYME, PH.D.   Professor of Epidemiology, Department of Biomedical and Environmental Health Sciences, School of Public Health, University of California, Berkeley, California 94720

STEPHEN WEISS, PH.D.   Chief, Behavioral Medicine Branch, National Heart, Lung & Blood Institute, Federal Building, Room 604, 7550 Wisconsin Avenue, Bethesda, Maryland 20205

DR. REDFORD B. WILLIAMS, M.D.   Professor, Duke University Medical Center, Department of Psychiatry, Box 3416, Durham, North Carolina 27710

Preface

During the past decade, cardiovascular medicine has made significant strides in the diagnosis and treatment of coronary heart disease and related disorders as well as developing a better understanding of potential preventive, risk-reducing measures. Highly sophisticated diagnostic instrumentation, surgical procedures, and emergency medical care have undoubtedly contributed greatly to the trend of declining mortality from cardiovascular events. In the course of the extensive research efforts associated with this area, it has become increasingly apparent that the identified coronary risk factors share the common element of *lifestyle* as a major determiner of health behaviors associated with these factors. Further, it is suspected that behavioral and environmental factors may play a significant contributory role toward the etiology and progression of arteriosclerosis through their effects on the central and peripheral nervous systems and associated neurohormonal response.

Considerable research effort has been devoted to identifying the risks associated with smoking, elevated blood pressure, and serum cholesterol. Research development and modification is being vigorously pursued. Behavioral research exploring the role of psychological stress factors, social support networks, stressful life events, and the Type A/B behavior patterns has uncovered potentially meaningful associations between behavioral factors and arteriosclerosis, as discussed in the succeeding chapters of this book.

Continued progress toward ultimately identifying both the most effective behavioral health approaches as well as toward uncovering biobehavioral "mechanism" linkages will be limited by two factors: 1) the development of conceptual models which adopt an *integrative* framework in assessing the relative contributions of behavioral, neural, genetic, dietary constitutional, and developmental factors to disease (as well as the synergistic or catalytic potential resulting from interactions of the above); and 2) the development of training opportunities necessary to prepare a cadre of scientists capable of taking full advantage of such multidisciplinary approaches toward understanding the mechanism of action as well as the techniques to treat and prevent cardiovascular disease.

This volume will attempt to address these issues by examining the various lines of evidence linking behavioral and environmental factors to the development, progression, treatment, and prevention of arteriosclerosis. Following the introduction, *Behavior and Arteriosclerosis* is organized into Parts: (I) biobehavioral research relating to arteriosclerosis; (II) the basic arteriosclerosis disease process; (III) clinical manifestations and management; (IV) prevention of arteriosclerosis; (V) research training; and (VI) overview and synthesis.

Part I, biobehavioral research, describes some of the findings of this new interdisciplinary science that combines basic, clinical, and epidemiologic research relating to arteriosclerosis.

Part II, the basic arteriosclerotic disease process, begins with consideration of the neurochemical mechanisms whereby somatomotor behavior and associated physiological processes are influenced by environmental events. The discovery of neurochemical substances and neurophysiological processes that mediate pain, subjective responses, and neuroendocrine activities has provided links between environmental events, brain biochemistry, and behavior.

The section continues with a consideration of the behavioral and physiological responses elicited by environmental events. Studies of behavior under controlled laboratory conditions have revealed psychological mechanisms that determine the behavioral and physiological responses to stressful, challenging situations.

Next, Part II addresses the sociocultural factors associated with arteriosclerotic cardiovascular disease. The relationship of social and cultural mobility, type A-coronary-prone behavior patterns, and stressful life events with coronary heart disease suggests that interrupted social ties predispose individuals to arteriosclerotic cardiovascular disease. Prospects for future research include studies of the physiological mechanisms linking behaviors elicited by social factors to pathogenesis of arteriosclerosis.

Finally, Part II discusses the pathogenesis of complications associated with arteriosclerotic cardiovascular disease. The morbid consequences of this disease are influenced not only by site and extent of lesions but also by neuroendocrine mechanisms affecting cardiac rhythm, myocardial oxygen requirements, platelet aggregation, and thrombosis.

Part III, clinical manifestations and management, begins with principles of behavior analysis. Experimental and clinical studies have revealed basic determinants of behavior, and many research programs currently are concerned with the application of these principles to cardiovascular risk factor reduction. Most attention has been given to the role of reinforcement schedules and stimulus control factors in initiating desired behaviors. Future research will reveal the role of these factors in maintaining behaviors for prevention of arteriosclerosis over long periods of time.

The section continues with a discussion of patient characteristics that influence outcomes of medical and surgical treatments. Just as clinical and hemodynamic data can be used to predict the outcome of treatments, so can behavioral and psychosocial variables be used to predict the benefits of medical and surgical therapies. The need to identify predictors of pain relief with surgical management is especially salient at the present time.

Programs of cardiac rehabilitation are studying the interactions between physical and psychological processes. Research results indicate that clinical responses of patients recovering from myocardial infarction are influenced by behavioral as well as physiological recovery processes. Much more information is needed concerning the correlates of cardiovascular function and the behavioral consequences of severe cardiac disease. More attention needs to be directed to methods for improving adherence to medical regimens for reducing risk factors and increasing capacity for physical work.

Part IV, prevention of arteriosclerosis, begins with an analysis of psychosocial processes influencing behavior. Most of the processes related to risk factors begin in childhood, and their development can be understood only by the study of children and adolescents. A basic assumption for our efforts to prevent arteriosclerosis is that risk factors are more capable of being prevented in children than of being reversed in adults. This assumption must be tested rigorously. In the meantime, it is apparent that once risk factors are established in adults, intensive psychological, social, and educational efforts are required to reverse them.

Part IV continues with a report of the results from comprehensive community programs for control of cardiovascular diseases in this country and in Europe. The main objective of these programs has been to reduce morbidity and mortality from cardiovascular disease and to reduce risk factors for arteriosclerosis. Even though enormous efforts must be made to reduce risk factors in the community, some progress has been made. In addition, the recent reduction in mortality from coronary heart disease in this country may be the result of sociocultural influences occurring spontaneously; our approaches to prevention seek to harness the relevant social, psychological, and behavioral forces.

Prospects for biobehavioral research in the future depend on building our resources for this research. Interdisciplinary research depends on training young investigators in more than one field and supporting their research activities. Because of the need for scientists with interdisciplinary experience, Part V recommends instituting programs for training scientists in the psychological and behavioral aspects of arteriosclerosis. The record shows that the National Heart, Lung and Blood Institute has provided leadership in developing these resources for biobehavioral research.

Part VI attempts to synthesize the issues raised and to suggest prospects for future biobehavioral research and research training in arteriosclerosis.

The impetus for this book came from several sources. We are indebted to many senior investigators involved in various Arteriosclerosis Specialized Centers of Research around the country, to the Working Group on Arteriosclerosis of the National Heart, Lung and Blood Institute, to the Committee on Human

Resources of the National Research Council, and to the many behavioral and biomedical scientists who have devoted major portions of their scientific careers to investigating the issues described herein.

Collectively, these sources of inspiration identified the basic concerns; our efforts were directed, therefore, toward collation and synthesis into a single volume to serve as a reference to the growing biobehavioral science community. If the past ten years of activity presage the future, the next decade should witness new insights into our understanding of the dynamics and interplay of biobehavioral factors in health and illness. No greater contribution toward the extension and quality of life can be achieved than acquiring such knowledge in the prevention and control of the primary source of coronary heart disease: arteriosclerosis.

Contents

Introduction

Emergence of Behavioral Medicine

J. Alan Herd

The extent of scientific knowledge in behavioral medicine has increased remarkably during the past decade. In the early 1970s, the term behavioral medicine had not even been defined in any consistent way. Psychosomatic medicine occupied the area between behavioral and biomedical sciences with its emphasis on the psychological basis for organic diseases such as hypertension, peptic ulcer disease, and ulcerative colitis. The behavioral sciences, however, were well developed and many investigators trained in psychology, sociology, and ethology were developing an interest in applications to biomedical science.

In 1970, the science of clinical psychology concerned itself with diagnosis and treatment of mental illness, and the treatment of patients with emotional disturbances. Clinical psychologists had developed an effective approach to treatment of patients through psychotherapy of individual patients and patients in groups. Analysis of psychotherapeutic processes was concerned with techniques for improving function in patients with psychotic disorders and severe behavioral disturbances. In addition, the techniques of behavior therapy and behavior modification had become well-established. These were applied most effectively to treatment of abnormal behaviors in patients who perceived their need for treatment.

The science of social psychology embraced such areas of research as attitudes and opinions, communication theory, and principles of social learning. In particular, statistical analyses of behaviors observed in social settings and results of measurements using psychometric instruments formed the basis for the principles of social psychology. Influences of social factors on education, motivation, and learning, and social determinants of behavior were well developed areas of research.

Psychological studies of infants, children, adolescents, and adults emphasized the transition of psychological factors during maturation and socializa-

J. Alan Herd • Sid W. Richardson Institute for Preventive Medicine, The Methodist Hospital, 6565 Fannin, MS S400, Houston, Texas.

tion. Topics of concern included learning, cognitive development, language and intelligence, perceptual and motor development, emotional development, personality, and socialization. The psychological mechanisms operative at each stage had been identified, and their impact on normal and abnormal function was under research.

Study of animal behavior had produced basic information concerning instrumental and operant conditioning, neurophysiology of behavior, and ethology. Several animal species including nonhuman primates had been studied extensively, and a basic knowledge of comparative psychology was in hand.

Although few biomedical scientists in 1970 were conducting research in coronary risk factor reduction, behavioral scientists were applying behavioral techniques in biomedical research. The treatment of obesity had been undertaken using operant techniques, self-management techniques, psychotherapy, and behavior modification. The realization that cigarette smoking was maintained by psychological factors as well as by pharmacological effects of nicotine led to study of developmental, social, and pharmacological influences on smoking. Although the physiological effects of nicotine could be overcome readily, the psychological and environmental influences proved difficult to correct. Research on persuasion and attitude change had revealed that fear appeal has little effect on most people. Since the Surgeon General's Report linking cigarette smoking to lung cancer had been published, it had become evident that fear of cancer had little effect on peoples' cigarette smoking habits. More immediate consequences perceived in some tangible way had been noted to have greater effect on attitudes and behaviors than long-term, remote possibilities of adverse consequences. Accordingly, a variety of operant techniques, self-management techniques, and aversive conditioning techniques had been employed in an effort to reduce cigarette smoking.

The realization that patients with medical illnesses frequently failed to follow their medical regimens led to the analysis of social and psychological factors underlying patients' responses to medical treatment Of great concern was the failure of patients with arterial hypertension to continue taking antihypertensive medication to control levels of arterial blood pressure when the patients felt well. Additional concerns arose through difficulty in bringing patients with medical illness into treatment programs. In particular, efforts to restore patients with ischemic heart disease to normal function frequently were limited by psychological and social factors as well as by physical disabilities. The realization that a broad approach to rehabilitation was necessary brought behavioral science into cardiovascular rehabilitation programs.

The state of behavioral science in arteriosclerosis research was reviewed in the 1971 Report of the Task Force on Arteriosclerosis (1). The Task Force noted the importance of three major risk factors: arterial hypertension, hyperlipidemia, and cigarette smoking. The Task Force also noted other risk factors that had some lesser influence. These included extreme obesity, sedentary living, emotional stress, personality factors, certain behavior patterns, and dietary intake of total calories, saturated fats, and cholesterol. Many of these other risk factors included behavioral features, but there was no definitive information linking arteriosclerosis and behavioral factors. The realization that patients with arterial hypertension were influenced by behavioral factors was indicated by concern with studying patients who neglected taking adequate treatment for hypertension. It had been recognized that patients failed to adhere to long-term therapy because of social and psychological factors.

The goals for behavioral medicine expressed by the Task Force were recommendations for fundamental research into atherosclerosis and clinical research in a multifactor trial of intervention in men with cardiovascular risk factors. The major emphasis was on research programs for smoking avoidance, reduction of arterial hypertension, and diet modification. It was recommended that improved methods for intervention be developed in clinics that included behavioral scientists on their professional staffs. The role of behavioral science in controlling hyperlipidemia was noted through the need for dissemination of information concerning diet and weight control. Specific approaches were not known, however, and little insight had been gained concerning measures for diet modification. Psychologists, sociologists, and other behavioral research scientists were to be encouraged to study the problems associated with health education and attitudes, motivation, and compliance as they relate to the prevention, therapy, and rehabilitation of patients with coronary heart disease, hypertension, or stroke.

The Task Force also was concerned with the mechanisms whereby genetics, hypertension, cigarette smoking, diabetes, physical exercise, and behavior exert their effects. The Task Force recommended strong support of basic research in atherosclerosis and other aspects of the biological sciences since "important discoveries related to atherosclerosis might well be made in fields outside cardiovascular research."

A concern for prevention also was expressed strongly. The understanding of pathogenesis underlying atherosclerosis was to be linked with the capability for intervention in disease-prone individuals. For this, the Task Force proposed the development of programs in which environmental modifications are attempted at a community level utilizing mass media for educational purposes. Thus, the nature of interventions was to range from education to medication.

Additional status reports and recommendations were published during the 1970s by the 1977 Working Group on Arteriosclerosis (2) and the Fifth Report of the Director (NHLBI) 1977 (3). Concern with discovery of causes for atherosclerosis was expressed in recommendations for fundamental research. In addition to concerns for pathophysiological mechanisms, recommendations were made for study of mechanisms whereby cigarette smoking, diabetes, physical exercise, and behavior exert their effects. The development and application of behavioral intervention was recommended repeatedly. However, no mention was made of the social, psychological, or behavioral factors likely to influence health-related behaviors. The realization that research in rehabilitation would be necessary for improved medical outcome was expressed through recommendations that a comprehensive approach be taken. This approach was to include social and psychological considerations as well as purely medical issues. In general, it was recommended that information be developed on techniques that promote or inhibit the application of knowledge about atherosclerosis to its prevention, diagnosis, and treatment.

A recurring theme in recommendations from several sources was an expanded program of advice to the public regarding the relationship between smoking and heart disease, nutrition and heart disease, and the risk factors associated with obesity. In addition, a major effort was recommended concerning creation of an Office of Health Education to include high blood pressure education as well as education on the nature of atherosclerosis and other major cardiovascular risk factors.

1. State-of-the-Art

The past decade of research in biobehavioral science has indicated that behavioral processes are associated with etiology and pathogenesis of arteriosclerosis. In addition, they affect the major risk factors for arteriosclerotic cardiovascular disease. They influence the outcomes of medical and surgical treatments for arteriosclerotic cardiovascular disease, and they determine the efficacy of preventive measures. Results of this research have been summarized and reviewed in the 1981 Report of the Working Group on Arteriosclerosis (4).

Physiological responses which occur under many different environmental and behavioral conditions appear to influence the basic process of arteriosclerosis and may precipitate the complications of ischemic heart disease. These responses include vascular spasm, platelet aggregation, lipid mobilization, endocrine hypersecretion, and autonomic dysfunction. Examples of psychological and behavioral stimulation are resting and arousal responses, cognitive tasks such as mental arithmetic and reaction time tasks, interpersonal competition, coping, adaptation, and strong emotion. Individual differences are seen in qualitative and quantitative responses and may be predictive of pathological consequences.

An association between neuroendocrine processes and arteriosclerosis is suggested by several clinical and experimental studies. These studies have produced evidence indicating a correlation between plasma levels of cortisol and severity of coronary arteriosclerosis in asymptomatic male subjects who had coronary angiography as part of a medical evaluation. Studies in nonhuman primates indicated that cortisol administered orally to monkeys fed a high-cholesterol diet increased involvement of the aortic intimal surface area with atherosclerotic lesions compared to animals receiving a high-cholesterol diet only. Numerous other studies have been undertaken to determine the relation between adrenocortical function and psychological distress. In general, the adrenal cortex has been found to be stimulated under environmental conditions perceived by a subject as threatening. These conditions produce adrenocortical responses not only when physical harm is threatened but also during psychological distress not associated with threat of physical injury.

Central nervous system structures have been identified that integrate autonomic and endocrine outflows, which, in turn, influence the cardiovascular system and may contribute to arteriosclerosis. These include hypothalamic areas such as AV3V, ventral medulla, fastigial nucleus, nucleus tractus solitarius, the raphe complex, and the intermediolateral cell columns in the spinal cord. Anatomical and electrophysiological techniques provide the momentum for studying the interactions among these regions. In addition, transmitters already identified provide the basis for studies of communication among these areas. These transmitters include angiotensin, vasopressin, oxytocin, glutamate, gama-amino butyric acid, serotonin, and other neural peptides. Although this basic knowledge is only beginning to provide an understanding of the basic neurobiology involved in the central control of cardiovascular function, results indicate that techniques are available to develop a more complete understanding of these mechanisms.

The Working Group concluded that certain patterns of behavior seem to enhance the prospect for developing clinical manifestations of atherosclerosis. These patterns of behavior have been seen in individuals who respond to challenges with great competitive drive, hostility, and impatience. These behavior patterns are called coronary-prone or Type A behavior patterns.

Numerous epidemiologic studies have demonstrated social and psychological factors associated with coronary heart disease. There is evidence that stressful life events are likely to be followed by a wide variety of illnesses including clinical ischemic heart disease. The impact of stressful life events appears to be modulated by the presence or absence of social support networks in the individual's environment.

The Working Group also concluded that principles of conditioning and social learning are important in determining habits of diet, smoking, and exercise. Principles of learning have been utilized in developing new approaches to cardiovascular risk factors such as cigarette smoking and obesity. Principles of social learning also have been used successfully in reducing cardiovascular risk factors. Behavioral and sociocultural factors also have been demonstrated to influence clinical management and rehabilitation of patients with arteriosclerotic cardiovascular disease. Characteristics which identify subgroups with greater likelihood of survival or successful treatment have been used to make rational choices of medical or surgical interventions. Behavioral interventions have been used to influence characteristics likely to influence outcomes of treatment.

Techniques for behavior modification have been developed for control of obesity and for smoking cessation. Many of these studies have been well controlled and patients have been followed for many months or years to determine long-term results. Efforts have also been made to facilitate the majority of smokers who want to quit on their own with minimal outside aid. In addition, treatments have been developed for special groups of smokers including pregnant women, blue-collar workers, adolescent smokers, and patients with medical disease adversely affected by cigarette smoking. The social base of health and health-related behaviors has been used to aid in smoking cessation, weight control, and long-term adherence to antihypertensive medication. Social support systems are known to facilitate development of coping strategies that help contain psychological distress within tolerable limits. It is plausible that social support networks may exert their influences on health and disease largely through neuroendocrine mechanisms.

Investigators have studied the exploratory, risk-taking behavior of adolescents. The years of adolescence are highly formative for health-relevant behavior patterns such as smoking cigarettes, food intake, and patterns of exercise. Advances in endocrinology, psychology, and epidemiology have produced information concerning adolescent development during the critical period of biological and psychological change.

Advances also have been made in the study of animal models. Several species of animals including nonhuman primates have been studied under classical and operant conditioning techniques and in social situations. For example, studies of behavioral procedures and social influences on behavior have been carried out in nonhuman primates receiving atherogenic diets. Measures of hemodynamics, and endocrine and metabolic parameters have been carried out concurrently with behavioral and dietary interventions. In these studies, scientists have developed techniques to produce behavioral effects through manipulation of environmental and social factors, and have observed systematic changes in behavior and cardiovascular function. Researchers have also observed pathological consequences of these interventions.

Finally, several health education and community multiple-measure programs concerned with cigarette smoking, levels of serum cholesterol, and elevations of blood pressure have reduced these risk factors in large populations. A few studies conducted with large populations over periods of several years have shown associations between reduced prevalence and severity of cardiovascular risk factors and reduced incidence of arteriosclerotic cardiovascular disease.

2. Summary

Although the neurobiological links between behavioral processes and arteriosclerosis are still unknown, a pattern is emerging. Physiological and psychological stimuli that are aversive modify both adrenomedullary and adrenocortical activity through secretion of neuropeptides as well as by direct neural influence. These neuropeptides also modify behavior and may be conditioned by environmental stimuli to be secreted even in the absence of painful stimuli. Adrenomedullary and adrenocortical secretions appear to influence the development of arteriosclerosis and precipitation of complications. However, the pathophysiological mechanisms involved are still largely unknown.

Results of clinical research indicate that behavioral factors are involved in medical, surgical, and pharmacological approaches that influence cardiovascular risk factors, adherence to medical regimens, and outcomes of treatment. Principles of conditioning and social learning have been demonstrated to influence behaviors which affect cardiovascular risk factors. Behavioral approaches to weight control have demonstrated several techniques for achieving short-term weight reduction. However, satisfactory techniques for maintaining weight reduction over long periods of time have not yet been developed. Research on smoking cessation shows that several technqiues can produce 15–25% one-year abstinence rates. A critical problem in smoking cessation is maintenance of nonsmoking behavior over long periods of time. Studies of factors influencing adherence to antihypertensive regimens indicate the important role of patient–clinician interaction and social support of the patient. As a result of these studies, it is apparent that once risk factors and health-damaging behaviors are established in adults, intensive psychological, social, and educational efforts are required to reverse them. Risk factors related to health behaviors are therefore more likely to be prevented in children than reversed in adults.

Prospects for biobehavioral research in the future depend upon building our resources for this research. Interdisciplinary research depends upon training young investigators in more than one field and supporting their research activities. Accordingly, we must train scientists in the psychological, social, and behavioral aspects of arteriosclerosis as well as in the biomedical approach to arteriosclerosis research.

References

1. National Heart and Lung Institute Task Force on Arteriosclerosis, National Institutes of Health. DHEW Publication No. (NIH) 72-219, Volume II, 1971.
2. Arteriosclerosis. The Report of the 1977 Working Group to Review the 1971 Report by the National Heart and Lung Institute Task Force on Arteriosclerosis. DHEW Publication No. (NIH) 78-1526, 1977.
3. Levy, R. I. Heart, lung, and blood research. Five years of progress: The challenge ahead. DHEW Publication No. (NIH) 78-1415, 1978.
4. Working Group on Arteriosclerosis of the National Heart, Lung, and Blood Institute. Arteriosclerosis, 1981; summary, conclusions, and recommendations. NIH Publication Nos. 81-2034 and 81-2035, Volumes 1 and 2, 1981.

Part I

Biobehavioral Research Relating to Arteriosclerosis

Chapter 1

Biobehavioral Research Relating to Arteriosclerosis

J. Alan Herd

1. Introduction

The influence of behavioral processes on arteriosclerosis is evident in pathogenesis of the basic process, management of clinical manifestations, and prevention of arteriosclerotic cardiovascular disease. Recent progress in all these areas has come from interdisciplinary research in which the interactions between physiological and behavioral processes have been explored. Results of this collaboration have come to be known as biobehavioral science, an interdisciplinary science that includes basic, clinical, and epidemiologic research. Basic neurobiologic studies have indicated mechanisms whereby behavioral processes influence metabolic and hemodynamic processes. Principles of behavior analysis have revealed basic determinants of human behavior in reduction of cardiovascular risk factors. Developmental approaches toward behaviors related to cardiovascular disease have achieved some success in reducing incidence of addictive cigarette smoking in youngsters. Finally, comprehensive community programs for control of cardiovascular risk factors have been successful in reducing incidence of cardiovascular disease. Thus, the record shows that a substantial proportion of our recent progress in basic knowledge, prevention, and management of arteriosclerotic cardiovascular disease has been the result of progress in biobehavioral science.

2. Neurobiologic Mechanisms in Arteriosclerosis

Associations between neuroendocrine functions and behavior have been known for some time. Recently, evidence has been obtained for associations

J. Alan Herd • Sid W. Richardson Institute for Preventive Medicine, The Methodist Hospital, 6565 Fannin, MS S400, Houston, Texas 77030.

between neuroendocrine function and arteriosclerosis. Finally, results of recent research have demonstrated neurochemical processes involved in the interaction of behavior and neuroendocrine function. It now seems likely that the physiological and neurochemical mechanisms mediating effects of behavioral processes on arteriosclerosis will become evident in the near future.

Basic neurobiologic studies indicate that the mechanisms whereby psychological, social, and behavioral processes influence metabolic processes apparently reside in the hypothalamus and other basal centers of the brain. New methods for tracing neuroanatomical circuits have been able to specify the routes taken within the central and peripheral nervous system and to specify many of the neurotransmitters involved in integration of endocrine and autonomic nervous system function. The links evident between behavioral and neurobiological systems reveal metabolic and endocrine influences on central nervous system function and environmental influences on central neurochemical processes as well as central control over regulatory systems and central influences on behavior. Recent evidence indicates that physiological variables are monitored by the central nervous system, which in turn controls appetitive and physiological factors that may contribute to development of obesity, diabetes mellitus, and other metabolic processes contributing to arteriosclerosis.

Epidemiologic studies have demonstrated several psychological, social, and behavioral factors associated with coronary heart disease. Psychological events that have been linked to coronary heart disease include dissatisfaction with interpersonal relations, job demands beyond a person's control, and a sudden, severe sense of loss. Sociocultural factors linked to coronary heart disease include major social change and interrupted social ties. Behavior patterns associated with coronary heart disease are characterized by great competitive drive, hostility, and impatience. The psychological and physiological processes underlying these increased risks for coronary heart disease involve responses of subjects to environmental challenges or loss of control over environmental situations. The effects of these psychological, social, and behavioral risk factors on coronary heart disease apparently are approximately as great as other risk factors such as hypertension and cigarette smoking.

Additional effects of psychological, social, and behavioral factors operate through other risk factors such as cigarette smoking, obesity, and physical inactivity. Studies of human subjects and experimental animals reveal that normal behavior is governed by principles of conditioning and reinforcement that are influenced strongly by social factors as well as by physiological factors. Furthermore, education alone is not enough to change high-risk health behaviors. Instructional control of behavior is relatively weak unless it is combined with more salient effects of physiological, psychological, or social reinforcements. Empirically-derived relations between environmental conditions, physiological conditions, psychological processes, and social variables have a profound effect on risk factor reduction and adherence to medical regimens.

The developmental biology of adolescence is particularly relevant for control of risk factors. The dramatic neuroendocrine changes of puberty are associated with drastic psychological and social transitions. This period is also a time of exploring behavior patterns that may damage health such as cigarette smoking, poor eating habits, and physical inactivity. Coronary-prone behavior patterns also emerge during adolescence. Field studies of social factors reveal the effects of various social processes such as peer influence and modelling as well as physiological effects on high-risk behaviors.

3. Biobehavioral Factors in the Clinical Management of Coronary Heart Disease

Clinical experience has shown that outcomes from treatment of coronary heart disease are affected by clinical, hemodynamic, and anatomic characteristics. Recent experience also has shown that outcomes are affected by psychological, social, and behavioral characteristics. For example, marital status is a strong predictor of survival in patients with coronary heart disease, and high intensities of hysteria and hypochondriasis decrease the likelihood of relieving pain of angina pectoris by either medical or surgical treatment. These relations between psychological, social, and behavioral characteristics and outcomes of treatment are independent of hemodynamic function, left ventricular contraction patterns, and extent of lesions shown by coronary arteriograms.

The need to identify predictors of pain relief with surgical management is especially important at the present time. Most patients with high hysteria scores and high hypochrondriasis scores on psychological tests are unlikely to obtain pain relief with medical management. However, surgical treatment of patients with these psychological characteristics and similar anatomic and hemodynamic characteristics produces pain relief in most of them.

Although the size and severity of a myocardial infarction are important determinants of outcome, psychological, social, and behavioral characteristics also are important in determining outcome of treatment for acute myocardial infarction. Individuals who become depressed following acute myocardial infarction have poor social readjustments, often fail to return to work, and have greater requirements for medical care. Individuals with coronary-prone behavior patterns (Type-A) have a greater possibility for reinfarction. The presence of either depression or coronary-prone behavior patterns are risk factors for complications that are quantitatively as great as hypertension or hyperlipidemia.

Psychological, social, and behavioral factors also affect adherence to prescribed medical regimens. Noncompliers tend to have coronary-prone behavior patterns, to have had previous myocardial infarctions, and to smoke cigarettes. Also, individuals who are depressed are less likely to comply. When patients will adhere to cardiac rehabilitation programs involving endurance exercise training and reduction of high-risk behaviors such as cigarette smoking, they return to work sooner and have fewer requirements for ongoing medical care.

The search for links between coronary arteriosclerosis and clinical events is based upon the inconsistencies between severity of coronary artery disease and likelihood for sudden cardiac death. As part of this search, the role of autonomic nervous system function and neuroendocrine processes has been examined, and links with sudden cardiac death have been found. Evidence that ergonovine or epinephrine plus propranolol can induce spasm in susceptible patients suggests that neuroendocrine processes may be involved in myocardial ischemia. Evidence for increased platelet adhesiveness and increased platelet growth factor has been observed in association with increased plasma levels of catecholamines. Injecting large amounts of catecholamine in experimental animals causes disruption of myocardial fibers and severe disturbances of cardiac rhythm. The pathophysiological processes that initiate these cardiac events consistently point to interactions between the neuroendocrine system and other biochemical and physiological systems.

4. Human Behavior and Cardiovascular Risk Factors

Success in management of patients with arteriosclerotic cardiovascular disease and in prevention of arteriosclerosis depends in part on our understanding of human behavior. Although there have been many explanations of human behavior and many efforts made to direct individuals to behave in certain ways, the scientific basis for explaining and directing behavior rests most firmly on the principles of the type of learning known as operant conditioning, which is based upon antecedent stimuli and the effects of consequences which follow behavior. In simplistic terms, behavior analysis is concerned with identifying and controlling the stimuli, rewards, and punishment that influence behavior. As the psychology of learning has matured, our understanding of human behavior has grown to include cognitive processes, interactions between individuals and the environment, and the role of social influences on behavior. In all these approaches to understanding human behavior, the emphasis has been on observations of overt events rather than analysis of subjective factors.

The scientific foundation for contemporary studies of human behavior developed from research on relations between physiology and behavior. The application of objective methods of physiology to the study of behavior began with Russian physiologists in the second half of the nineteenth century. Known as conditioning, this theory developed concepts about the reflexive nature of behavior. Ivan M. Sechenov (1965) combined experimental studies of cerebral influences on spinal reflexes with theoretical analyses to conclude that all behavior was entirely reflexive. Ivan P. Pavlov (1927) applied physiological methods to studies of conditional reflexes. He investigated several processes associated with conditioning and described processes leading to acquisition and extinction of conditional reflexes. Vladimir M. Bechterev (1933) also used objective methods to demonstrate conditioning and extended principles of conditioning to studies of behavioral and psychiatric disorders. All three physiologists proposed hypotheses about the role of the nervous system in learning and used objective physiological methods to study psychological processes.

Application of objectivism in psychology continued during the twentieth century in the work of American psychologists. Although physiological techniques were not used by all investigators, their experimental work in animal psychology was based upon purely objective techniques. John B. Watson (1919) advocated the study of various environmental stimuli and the responses they evoked. Edward L. Thorndike (1933) used laboratory methods of research to study acquisition of new responses and developed laws of learning based upon effects of rewards. The experimental works of these investigators established the objective approach to the study of behavior. These investigators also demonstrated the complexity of the learning process.

The experimental analysis of behavior reached its present level of importance through the work of B. F. Skinner (1938). He distinguished two types of conditioning according to the type of response involved (1935). The process of eliciting responses through reflexes caused by stimuli such as food or electric shock was called classical conditioning, while developing responses through presentation of stimuli contingent upon those responses was called operant conditioning. Skinner's distinction between these two types of conditioning was a major contribution because elicited or reflex responses could not account for most normal behavior. He elaborated the basic principles of operant conditioning and identified variables that contribute to operant behavior. He also developed methods for studying behavior in laboratory animals using automated equipment to present stimuli and record responses.

The basic principles of operant conditioning are concerned with relations between behavior and the environmental events that influence it. These principles include reinforcement, punishment, extinction, and stimulus control. All of these principles have been studied in experimental animals and human subjects under laboratory conditions. They have also been demonstrated in human subjects under natural conditions outside the laboratory.

The principle of reinforcement refers to the effect of a stimulus or event occurring after some behavior to increase or decrease the likelihood the response will be repeated. Positive reinforcers are stimuli or events occurring after a response that increase the frequency of the response. An example of a positive reinforcer is delivery of food to a hungry animal after that animal has pressed a lever a predetermined number of times in a specified period of time. Negative reinforcers are stimuli or events that increase frequency of response after the reinforcer is removed. An example of a negative reinforcer is electric shock to an animal's skin that can be turned off or prevented if the animal presses a lever in a specified way. In these examples, the response of pressing a lever increases when food is delivered, or electric shock is terminated.

Reinforcement influencing human behavior under clinical conditions has also been demonstrated. Application of operant conditioning to human behavior was begun by Azrin and Lindsley (1956). They demonstrated that giving candy to children could be used to develop complex cooperative behaviors. Further extensions of operant conditioning have been made to clinical problems such as training of retarded children and caring for mentally disturbed adults. For example, Ayllon and Azrin (1968) used a system of tokens that could be redeemed for desirable things to teach psychotic patients various tasks and self-care activities. Since that time, operant techniques have been used to treat many different clinical disorders of behavior.

Reinforcers influencing normal human behavior under natural conditions are exceedingly complex. Ferster, Nurnberger, and Levitt (1962) noted that positively-reinforcing aspects of eating occur with ingestion of food, whereas overeating causes obesity, which has negatively reinforcing qualities. However, the adverse consequences of overeating are delayed and have less impact than the pleasure of eating. Further complexity arises from the influence of social stimuli. Studies by Evans, Rozelle, Mittelmark, Hansen, Bane, and Havis (1978) with a large population of seventh grade students suggest that peer pressure, models of smoking parents, and exposure to mass media influence the tendency of children to start smoking cigaretttes. Positive reinforcements under natural conditions apparently arise through highly complex processes.

The principle of punishment refers to the delivering or withholding of some stimulus or event after some response to decrease the likelihood the response will be repeated. An example of an aversive stimulus acting as a punisher is application of an electric shock to an animal's skin when that animal presses a lever, and the shock causes the rate of lever-pressing to decrease. An example of a positive stimulus used as a punisher is withholding delivery of food from an animal when that animal presses a lever, and the withholding of food causes a decrease in its rate of lever pressing. In these examples, the response of pressing a lever decreases when electric shock is delivered or food is withheld.

Punishments designed to influence human behavior abound in ordinary experience. However, the effects of punishment are complex. Laboratory experiments (Morse and Kelleher, 1977) have demonstrated that aversive stimuli may decrease response rates under some conditions and increase rates of responding under others. A practical example of the complex effects from punishment is seen in our criminal justice system in which criminals punished by imprisonment often repeat their criminal behaviors when they are released and come under the influence of environmental conditions that produced the criminal behavior in the first place. In laboratory experiments, the most effective use of punishment comes when the original behavior is not well maintained.

The principle of extinction refers to reducing the likelihood of a learned response occurring by withholding reinforcers that previously maintained that response. An example of extinction can be seen when food is no longer delivered to a hungry animal when it presses a lever. Extinction may also occur when food is delivered to an animal until it is satiated. In both examples, the tendency for the animal to press a lever is reduced. Here again, extinction occurs commonly in ordinary experience. Adults stop working when their pay is stopped, and children stop their childhood games when their peers adopt more adult activity. A person who overeats or smokes cigarettes may stop those high risk behaviors if the immediate pleasures are removed. However, operant behaviors in laboratory animals can be maintained for incredibly long periods of time even in the absence of primary reinforcers.

The principle of stimulus control refers to the effects of stimuli or events occurring before some behavior to increase or decrease the likelihood that the response will occur. Responses positively reinforced in the presence of particular enviornmental conditions tend to occur more frequently under those conditions than in their absence. For example, if food is delivered only in the presence of a light, a hungry animal will learn to press a lever to obtain food only when the light is on. When the animal responds differently to different stimuli, it is said to be under stimulus control. In this situation, the animal is likely to respond in a highly selective manner. Stimuli associated with reinforcers also come to acquire reinforcing properties in their own right. For example, an animal will actually respond to produce the stimulus even though the primary reinforcer is long delayed. By judicious scheduling of stimuli, it is possible to develop long sequences of behavior with delivery of primary reinforcers occurring only infrequently. The significance of this general principle of stimulus control lies in the complex interaction of primary and secondary reinforcers in maintaining highly complex behavior.

Stimulus control of human behavior is obvious under ordinary conditions. Some behaviors are developed from practical experience with stimuli, responses, and reinforcers. Many other behaviors occur because of instruction, practical example, and anticipation of likely future events. For example, a driver of an automobile will stop at a red light because of a belief that it is the wise thing to do rather than because that driver once had an accident or received a traffic ticket after ignoring a red light. Control of eating behavior and cigarette smoking also comes under stimulus control in that the stimuli that often elicit these complex activities have been associated with primary reinforcers in the past. One of the prerequisites for altering human behavior is to identify and modify environmental conditions associated with those behaviors.

Progress in research on health related behaviors has demonstrated the efficacy of behavioral intervention. These interventions have been tested in weight control and smoking cessation as well as in promoting adherence to medical regimens. Although some progress has been made, more research is necessary to bring efficacy of treatment for health related behaviors up to the efficacy observed for behavior therapy in general.

Behavioral treatment for weight control has been studied extensively. In general, treatment effects have been modest with few reports of sustained weight loss. Reports by Stuart (1967) indicated 80% of a group of obese subjects lost more than 20 pounds and 30% lost more than 40 pounds. Results, however, for the majority of studies in weight control have been less dramatic.

Intensive behavioral treatment for weight control has been reported to be most successful. Musante and colleagues at Duke University (1976) provided three meals a day to obese subjects in a clinic dining room with cognitive retraining of eating habits and practical experience under supervision. Of those obese subjects who stayed in treatment for six to eleven months, 85% lost more than 20 pounds and 61.5% lost more than 40 pounds. The long-term effects of this treatment remain to be seen.

Behavioral treatment for smoking cessation has many of the problems encountered in treatment for weight control. A variety of techniques have been effective in producing cessation in the short term, but few are effective in producing permanent cessation. Lando (1977) has reported 6-month abstinence rates of 76% in a group of smokers treated with rapid smoking aversion therapy. However, McAllister, Farquhar, Thoresen, and Maccoby (1976) have reported that only 30% of treated subjects become nonsmokers using any technique, while about 20% of chronic smokers are able to quit on their own.

Our challenge for the future is the application of basic principles of behavior to management of patients with arteriosclerotic cardiovascular disease and to prevention of arteriosclerosis. High-risk behaviors are common even in patients with clinical manifestations of coronary heart disease and in individuals with high levels of risk factors for cardiovascular arteriosclerotic disease. Our ability to alter high-risk behaviors depends on our understanding of conditioning and learning in human subjects and our analysis of antecedent, concurrent, and consequent events associated with behavior.

Behavior that influences cardiovascular risk factors may, in turn, be influenced by social factors. These social and cultural factors include influences of peers, parents, and teachers through instruction and example, reward, and punishment, which have reinforcing properties. Some of these influences compete with one another in affecting behavior. For example, effects from information about dangers of behavior such as cigarette smoking are in competition with observations of cigarette smoking in parents, teachers, and peers. Also, the glamour depicted in advertising for cigarette smoking acts to overcome any fears of its dangers to health.

Several investigators have demonstrated that fear of harm does have some impact on behavior. For example, patients who have suffered an acute myocardial infarction frequently change their habits of eating and cigarette smoking. However, many return eventually to their previous patterns of behavior (Marston, 1970). Apparently, habits such as cigarette smoking that have been well-established are little affected by fear of harm from continuing in those habits.

The challenge of influencing cigarette smoking has resulted in efforts to prevent children from becoming cigarette smokers rather than persuading cigarette smokers to stop smoking. Studies by Evans (1976) suggest that adolescent smokers do believe that cigarette smoking is unhealthy. However, studies by several groups of investigators indicate that peer pressure, models of smoking parents, and messages in mass media outweigh the belief of children that smoking is dangerous (Evans *et al.*, 1978). Thus, information alone has little effect in deterring cigarette smoking. Apparently, adolescents must be trained to cope with social influences on smoking behavior rather than depending primarily on communicating information concerning the dangers of smoking.

5. Biobehavioral Approaches to the Prevention of Coronary Arteriosclerosis

The clinical management and prevention of cardiovascular risk factors are founded upon basic principles of conditioning and social learning. The technology for applying these principles arose out of studies with experimental animals, studies with normal human subjects under laboratory conditions, and studies with human patients displaying abnormal behavior under many conditions. These principles are applicable to risk factor reduction, particularly when conditioning procedures include reinforcements that are relevant for human subjects under natural conditions. In addition to learning by conditioning, learning occurs by instruction and by observing others. Also, behavior can be maintained by social reinforcements, such as peer pressure, that are difficult to control under natural conditions. Ultimately, the challenge for successful intervention is to harness reinforcements built into social support networks where people actually live under conditions that will maintain desired behaviors over long periods of time.

The reduction of risk factors in young people is particularly complex. Being made aware of dangers from high-risk behaviors such as cigarette smoking has little effect on behavior. Social influences have stronger effects than education. Most youngsters know that cigarette smoking is dangerous for health, but many respond to peer influences and begin experimental cigarette smoking in junior high school. Some success in reducing incidence of addictive cigarette smoking in youngsters has been achieved by showing them the physiological effects of smoking and socially acceptable ways to refuse the offer of a cigarette. The most effective approach to high-risk behaviors in each group of youngsters depends upon physiological, psychological, and social influences on behavior during various stages of adolescence.

Training adolescents to resist social pressures to begin smoking has been accomplished. Studies by McGuire (1974) suggested that specific instructions in ways to resist the influence of persuasive communications can be effective. Since teenagers are most likely to begin smoking during the junior high school years, Evans, Rozelle, Dill, Guthrie, Hanselka, Henderson, Hill, Maxwell, and Raines (1980) trained students in specific ways of coping with the immediate social influences to smoke. Videotapes depicting adolescents in common social situations were presented to illustrate ways of resisting social pressure to smoke. Tests of smoking behavior included measures of nicotine in saliva as well as self-reports. Follow-up studies indicated substantial reductions in smoking among students in the treatment groups compared to control groups.

These results suggest that such interventions may prove more useful in deterring smoking among junior high school students than instruction on the remote dangers of smoking. Instructions in social coping techniques apparently reduce incidence of cigarette smoking more effectively than fear alone. Sophisticated instructional programs that inoculate students against social pressures toward high-risk behaviors may be most effective as a means of reducing cardiovascular risk factors that are influenced by an individual's behavior.

A broader approach to social influences on behavior also has proven effective in reducing cardiovascular risk factors. Starting in 1972, a community program for control of cardiovascular disease was begun in North Karelia, Finland (Puska, Tuomilento, Salonon, Neittaanmake, Maki, Virtamo, Nissinen, Koskela, and Takalo, 1979, and Salonen, Puska, and Mustaniemi, 1979). The intervention goal was reduction of cardiovascular risk factors with the expectation that morbidity and mortality from cardiovascular disease also would be reduced. Because elevated risk factor levels in most of the population are influenced by behavior, it was assumed risk factors would be linked to social factors. Consequently, a comprehensive community program was initiated using mass media, general health education measures, provision of practical services, training of local personnel, environmental changes, and installation of communication and information systems. All these activities were supported enthusiastically by governmental officials and the community.

The program's social impact, cost, and effect on risk factors were evaluated by examining independent representative population samples in North Karelia and another matched reference area. More than 10,000 subjects were studied at the outset in 1972 and after five years in 1977. Using a multiple logistic function for computing effects of cigarette smoking, concentrations of serum cholesterol, and levels of arterial blood pressure, a significant reduction in risk of 17.4% was observed among men and 11.5% among women in North Karelia compared to the reference area. At the same time, there was a reduction in incidence rate of acute myocardial infarction among men of 16.7% and among women of 10.2%. Incidence of cerebrovascular accidents declined 12.7% among men and 35.5% among women. Mortality from all causes also declined. The costs of the program were small and matched well with the evident benefits for the community.

Although average reductions in cigarette smoking, serum cholesterol values, and levels of blood pressure were small, the impact of the program on cardiovascular disease was substantial. Results of this community-based intervention trial indicate the effectiveness of social networks in diffusing information, initiating behaviors, and maintaining reductions in cigarette smoking, serum cholesterol values, and levels of blood pressure.

6. Conclusions

1. Associations have been established indicating possible links between behavior, physiological processes, and arteriosclerotic cardiovascular disease. More information, however, is needed concerning the pathophysiological links. In particular, the genetic determinants and the developmental processes influencing behavior must be determined as well as the psychological and physiological processes whereby behavior influences arteriosclerosis.

2. Principles of conditioning and social learning have been demonstrated in behaviors that influence cardiovascular risk factors, adherence to medical regimens, and outcomes of treatment for cardiovascular disease. More information is needed concerning the physiological and genetic determinants of high-risk behaviors and the psychological, social, and physiological determinants of human behavior under natural conditions.

3. Education and broad community multiple-measure approaches to cigarette smoking, levels of serum cholesterol, and elevations of blood pressure have reduced these risk factors and reduced incidence rates of cardiovascular disease. However, more information is needed concerning the predisposing fac-

tors governing the impact of cardiovascular risk factors and risk factor reduction in individuals within communities exposed to broad general interventions.

References

Ayllon, T., and Azrin, N. H. *The token economy: A motivational system for therapy and reha-bilitation.* New York: Appleton-Century-Crofts, 1968.

Azrin, N. H., and Lindsley, O. R. The reinforcement of cooperation between children. Journal of Abormal Social Psychology, 1956, *52*, 100–102.

Bekhterev, V. M. In E. Murphy & W. Murphy (Eds.), *General principles of human reflexology: an introduction to the objective study of personality.* London: Jarrolds, 1933.

Evans, R. I. Smoking in children: developing a social psychological strategy of deterrence. *Journal of Preventive Medicine,* 1976, *5,* 122–127.

Evans, R. I., Rozelle, R. M., Dill, C. A., Guthrie, T. J., Hanselka, L. L., Henderson, A. H., Hill, P. C., Maxwell, S. E., and Raines, B. E. The Houston Project: focus on target-based filmed interventions. In *Symposium on deterrents of smoking in adolescents: evaluation of four social psychological strategies,* American Psychological Association: Montreal, Quebec, Canada, 1980.

Evans, R. I., Rozelle, R. M., Mittelmark, M. B., Hansen, W. B., Bane, A. L., and Havis, J. Deterring the onset of smoking in children: knowledge of immediate physiological effects and coping with peer pressure, media pressure, and parent modeling. *Journal of Applied Social Psychology,* 1978, *8,* 126–135.

Ferster, C. B., Nurnberger, J. I., and Levitt, E. B. The control of eating. *Journal of Mathematics,* 1962, *1,* 87–109.

Lando, H. A. Succesful treatment of smokers with a broad-spectrum behavioral approach. *Journal of Consulting and Clinical Psychology,* 1977, *45,* 361–366.

Marston, M. L. Compliance with medical regimes: a review of the literature. *Nursing Research,* 1970, *19,* 312–323.

McAlister, A. L., Farquhar, J. W., Thoresen, C. E., Maccoby, N. Behavioral science applied to cardiovascular health: progress and research needs in the modification of risk-taking habits in adult populations. *Health Education Monographs,* 1976, *4,* 45–74.

McGuire, W. J. Communication–persuasion models for drug education: Experimental findings. In M. Goodstadt (Ed.), *Research on methods and programs of drug education.* Addiction Research Foundation: Toronto, Ontario, Canada, 1974

Morse, W. H., and Kelleher, R. T. Determinants of reinforcement and punishment. In W. K. Honig and J. E. R. Staddon (Eds.), *Handbook of operant behavior.* Englewood Cliffs, New Jersey: Prentice-Hall, 1977.

Musante, G. J. The dietary rehabilitation clinic: Evaluative report of a behavioral dietary treatment of obesity. *Behavior Therapy,* 1976, *7,* 198–204.

Pavlov, I. P. In G. V. Anrep (Ed. and trans.), *Conditioned Reflexes: An investigation of the physiological activity of the cerebral cortex.* London: Oxford University Press, 1927.

Puska, P., Tuomilehto, J., Salonon, J., Neittaanmake, L., Maki, J., Virtamo, J., Nissinen, A., Koskela, K., and Takalo, T. Changes in coronary risk factors during comprehensive five-year community programme to control cardiovascular diseases (North Karelia project). *British Medical Journal,* 1979, *2,* 1173–1178.

Salonen, J. T., Puska, P., and Mustaniemi, H. Changes in morbidity and mortality during comprehensive community programme to control cardiovascular diseases during 1972–1977 in North Karelia. *British Medical Journal,* 1979, *2,* 1178–1183.

Sechenov, I. M. In S. Belski (trans.) and G. Gibbons (Ed.), *Reflexes of the brain: An attempt to establish the physiological basis of psychological processes.* Cambridge, Massachusetts: MIT Press, 1965.

Skinner, B. F. Two types of conditioned reflex and a pseudo type. *Journal of General Psychology,* 1935, *12,* 66–77.

Skinner, B. F. *The behavior of organisms: An experimental analysis.* New York: Appleton-Century, 1938.

Stuart, R. B. Behavior control over eating. *Behavior Research Therapy,* 1967, *5,* 357–365.

Thorndike, E. L. *An experimental study of rewards.* New York: Columbia University Teachers College, 1933.

Watson, J. B. *Psychology, from the standpoint of a behaviorist.* Philadelphia: Lippincott, 1919.

Part II

Etiology and Pathogenesis of Arteriosclerosis

Biobehavioral Mechanisms in the Etiology and Pathogenesis of Coronary Heart Disease

Redford B. Williams, Jr.

The contribution of the traditional risk factors for coronary heart disease—hypertension, elevated serum cholesterol, and cigarette smoking—is well-documented. Increasing evidence, however, has also implicated neurobiologic, psychosocial, and sociocultural factors as contributing to the etiology and pathogenesis of coronary heart disease (Jenkins, 1971; Jenkins, 1976).

There is little doubt that the transduction of various psychosocial phenomena into pathogenic physiological and neuroendocrine responses in the body is mediated by the brain. It is reassuring, therefore, that parallel to research documenting an association between various psychosocial factors and CHD has been an explosion of knowledge within the field of neuroscience. This will eventually help to explain basic mechanisms whereby psychosocial factors are transduced into physiologic and neuroendocrine processes leading to disease. It is beyond the scope of this overview to cover these exciting developments in the field of neuroscience. To indicate the possible future convergences between developments in neuroscience and behavioral medicine, two brief examples are offered.

A key physiological mechanism serving to maintain cardiovascular homeostasis is the baroreceptor reflex, whereby pressoreceptors in the carotid sinus and other areas monitor increases in blood pressure and transmit that information to the nucleus tractus solitarius. From this, ascending fibers convey the information to the hypothalamus, with the result that sympathetic outflow to the heart and vasculature is decreased, thereby "damping" any rise in blood pressure. Recent research has demonstrated mechanisms whereby stress can

Redford B. Williams Jr. • Duke University Medical Center, Department of Psychiatry, Box 3416, Durham, North Carolina 27710.

act to "turn off" the normal baroreceptor reflex, such that activation of the hypothalamic defence areas results in a situation where sympathetic outflow is not diminished by stimulation of peripheral baroreceptors (Wennergren, Kisander, and Oberg, 1976). Thus, stress can circumvent normal homeostatic mechanisms whereby potentially pathogenic physiological and neuroendocrine responses are limited.

A number of studies have implicated noradrenergic neurons originating in the locus coeruleus as playing a role in the regulation of cardiovascular function. Based upon earlier studies showing that a blood pressure fall after shock-induced fighting in rats is mediated by CNS catecholamine neurons, a recent study attempted to identify the specific neurons involved by placing bilateral locus coeruleus lesions in one group of rats and sham lesions (lowering of electrode without application of current) in another group, and then examining the blood pressure response of each group to the shock-induced fighting paradigm (Williams, Richardson and Eichelman, 1978). As in the earlier studies with global destruction of CNS catecholamine neurons with intracisternal injection of the neurotoxic drug, 6-hydroxydopamine, those rats with bilateral locus coeruleus lesions showed a blood pressure increase after fighting. In contrast, the control animals showed a blood pressure decrease. This study provides one example of how recently accumulated knowledge of CNS neurotransmitter systems can be utilized to identify brain mechanisms for transducing psychosocial stress into potentially pathogenic physiologic responses.

The strongest demonstration of a relationship between any psychosocial factor and CHD is that showing a prospective association between the Type-A behavior pattern and increased risk of CHD events. Type-A persons are characterized by high levels of achievement striving, impatience, and aggressive, hostile behavior. In a prospective study of over 3,000 middle-aged men who were healthy at intake, The Western Collaborative Group Study found that those who were Type A experienced about twice as many clinical CHD events over an 8½ year follow-up in comparison to their Type-B counterparts (Rosenman, Brand, Shultz, and Friedman, 1976). This prospective association between Type-A behavior pattern and increased CHD rates has also been confirmed in the Framingham study (Rosenman et al., 1976). In addition to increased rates of chinical CHD, there is evidence which suggests that arteriographically-documented coronary atherosclerosis is more severe among Type-A patients than Type-B patients (Blumenthal, Williams, and Kong, 1978).

A number of studies have been undertaken to identify psychophysiological and neuroendocrine mechanisms that might account for the increased CHD rates observed among Type-A persons (Dembroski, Weiss, Shields, Haynes, and Feinlich, 1978). When challenged to perform a variety of behavioral tasks, Type As showed greater increases in heart rate, blood pressure, and catecholamine secretion. Among those psychological characteristics which have been implicated in the tendency to display overt Type-A behavior under challenge are an increased need for control (Glass, 1977), increased self-involvement (Scherwitz, Berton, and Leventhal, 1978), and increased levels of hostility (Matthews, Glass, Rosenman and Bortner, 1977; Williams, Haney, Lee, Kong, Blumenthal, and Whalen, 1980).

The status of Type-A behavior pattern as a CHD risk factor was confirmed by an NHLBI-convened panel of behavioral and biomedical scientists who reviewed the available evidence concerning Type-A behavior and CHD (Dembroski, *et al.*, 1978). This panel concluded that a Type-A behavior pattern confers increased risk of developing clinically apparent CHD, and that this increased risk " . . . is over and above that imposed by age, systolic blood pressure, serum cholesterol, and smoking, and appears to be of the same order of magnitude as the relative risk associated with any of these other risk factors."

In addition to the evidence supporting the involvement of chronic psychosocial forces in the initiation and progression of arteriosclerotic cardiovascular disease, and in the precipitation of acute clinical events, there is also evidence suggesting that acute massive psychosocial stress can precipitate sudden death or potentially fatal arrhythmias even in the absence of significant coronary atherosclerosis (Eliot, Baroldi, and Leone, 1974; Lown, DeSilva, and Lenson, 1978). It is hypothesized that stresses such as suddenly learning of the death of a loved one or being physically attacked cause such a massive outpouring of catecholamines that actual coagulative myocytolisis occurs, with concomitant fatal myocardial infarction or the precipitation of a fatal arrhythmia.

There is evidence that stressful life events are likely to be followed by a wide variety of illnesses, including clinical CHD events. The impact of stressful life events appears to be modulated by the presence or absence of social support networks in the individual's current environment. It has been shown, for example, that the risk of CHD increases with major changes in place of residence (Syme, Hyman, and Enterline, 1964; Syme, Borhani, and Buechley, 1965), major changes in occupation (Syme *et al.*, 1964; Syme *et al.*, 1965; Kaplan, Cassel, Tyroler, Coroni, Kleinbaum, and Hames, 1971; Shekelle and Ostfeld, 1969), and the occurrence of discrepancies between the culture of upbringing and current sociocultural situation (Syme *et al.*, 1964; Syme *et al.*, 1965; Kaplan, Cassel, Tyroler, Coroni, Kleinbaum, and Hames, 1971; Shekelle and Ostfeld, 1969; Medalie, Kahn, Newfeld, Riss, Goldbourk, Perlstein, and Oron, 1973). It is not clear at present whether this increase in risk occurs as a result of mobility per se or as a result of personal characteristics which predispose certain individuals to become mobile. There is at least one study (Tyroler and Cassel, 1964) showing increased CHD rates among persons who were not mobile but who experienced changes in the situation in which they lived. Another theme in the life-change literature addresses the role of various kinds of losses which may influence health (Rahe, 1972). Despite problems with retrospective designs in earlier work, a number of studies are now available in which these problems are not at issue. Thus, a number of studies have shown an increased rate of CHD among widows and widowers shortly following the death of their spouses, and other studies have shown increases in blood pressure among those recently unemployed (Dohrenwend and Dohrenwend, 1974; Satariano and Syme, in press).

In a recent 9-year follow-up study of 6,928 persons, an increased mortality rate was observed among persons prospectively identified as having fewer friends and contacts with other persons (Berkman and Syme, 1979). This relationship was linear and was independent of health status at baseline, as well as of other such risk factors as obesity, cigarette smoking, physical inactivity, and other health habits. Other studies of CHD rates among Japanese migrants to Hawaii and California have also documented the increased risk of CHD resulting from relatively weak social-support networks. It was found that increased CHD rates among Japanese migrants living in California could not be accounted for by any differences in age, diet, serum cholesterol, blood pressure, or cigarette smoking (Marmot and Syme, 1976). However, it was found in this study that those who retained traditional Japanese lifestyles and associations were far less likely to suffer clinical CHD events, leading to the inference that high levels of social supports in traditional Japanese culture were somehow protective.

References

Berkman, L. F., and Syme, S. L. Social networks, host resistance and mortality: A nine-year follow-up study of Alameda County residents. *American Journal of Epidemiology*, 1979, *109*, 186–204.

Blumenthal, J. A., Williams, R. B., Kong, Y., *et al.* Type A behavior and angiographically documented coronary disease. *Circulation*, 1978, *58*, 634–639.

Dembroski, T. M., Weiss, S. M., Shields, J. L., Haynes, S. G., and Feinlich, M. *Coronary-prone behavior*. New York: Springer-Verlag, 1978.

Dohrenwend, B. S., and Dohrenwend, B. P. *Stressful life events: their nature and effects*. New York: Wiley-Interscience, 1974.

Eliot, R. S., Baroldi, G., and Leone, A. Necropsy studies in myocardial infarction with minimal or no coronary luminal reduction due to atherosclerosis. *Circulation*, 1974, *49*, 1127–1131.

Glass, D. C. *Behavior patterns, stress and coronary disease*. Hillsdale, New Jersey: Lawrence Erlbaum Associates, 1977.

Haynes, S. G., Feinleib, M., Levine, S., Scotch, N., and Kahnel, W. B. The relationship of psychosocial factors to coronary heart disease in the Framingham Study. II. Prevalence of coronary heart disease. *American Journal of Epidemiology*, 1978, *107*, 384–402.

Jenkins, C. D. Psychologic and social precursors of coronary disease. *New England Journal of Medicine*, 1971, *284*, 244–255; 207–217.

Jenkins, C. D. Recent evidence supporting psychologic and social risk factors for coronary heart disease. *New England Journal of Medicine*, 1976, *294*, 987–994.

Kaplan, B. H., Cassel, J. C., Tyroler, H. A., Cornoni, J. C., Kleinbaum, D. G., and Hames, C. G. Occupational mobility and coronary heart disease. *Archives of Internal Medicine*, 1971, *128*, 938–942.

Lown, B., DeSilva, R. A., and Lenson, R. Roles of psychologic stress and autonomic nervous system changes in provocation of ventricular premature complexes. *American Journal of Cardiology*, 1978, *41*, 979–985.

Marmot, M. G., and Syme, S. L. Acculturation and coronary heart disease in Japanese-Americans. *American Journal of Epidemiology*, 1976, *104*, 225–247.

Matthews, K. A., Glass, D. C., Rosenman, R. H., and Bortner, R. W. Competitive drive, pattern A, and coronary heart disease: A further analysis of some data from the Western Collaborative Group Study. *Journal of Chronic Disease*, 1977, *30*, 489–498.

Medalie, J. H., Kahn, H. A., Neufeld, H. N., Riss, E., Goldbourt, U., Perlstein, T., and Oron, D. Myocardial infarction over a five-year period: I. Prevalence, incidence and mortality experience. *Journal of Chronic Disease*, 1973, *26*, 63–84.

Rahe, R. Subjects' recent life changes and their near-future illness reports: A review. *Annals of Clinical Research*, 1972, *4*, 250–265.

Rosenman, R. H., Brand, R. J., Sholtz, R. I., and Friedman, M. Multivariate prediction of coronary heart disease during 8.5 year follow-up in the Western Collaborative Group Study. *American Journal of Cardiology*, 1976, *37*, 903–910.

Satariano, W., and Syme, S. L. Life change and disease: Coping with change. In J. L. McGaugh, S. B. Kiesler, and J. G. March (Eds.), *Biology and behavior of the elderly*. Washington, D.C.: National Academy of Sciences, in press.

Scherwitz, L., Berton, K., and Leventhal, H. Type A behavior, self-involvement, and cardiovascular response. *Psychosomatic Medicine*, 1978, *XL*, 593–609.

Shekelle, R. B., Ostfeld, A. M., and Paul, O. Social status and incidence of coronary heart disease. *Journal of Chronic Disease*, 1969, *22*, 281–294.

Syme, S. L., Borhani, N. O., and Buechley, R. W. Cultural mobility and coronary heart disease in an urban area. *American Journal of Epidemiology*, 1965, *82*, 334–346.

Syme, S. L., Hyman, M. M., and Enterline, P. E. Some social and cultural factors associated with the occurrence of coronary heart disease. *Journal of Chronic Disease*, 1964, *17*, 277–289.

Tyroler, J. A., and Cassel, J. Health consequences of culture change: II. Effect of urbanization on coronary heart mortality in rural residents. *Journal of Chronic Disease*, 1964, *17*, 167–177.

Wennergren, G., Lisander, B., and Oberg, B. Interaction between hypothalamic defense reaction and cardiac ventricular receptor reflexes. *Acta Physiological Scandinavica*, 1976, *96*, 523.

Williams, R. B., Haney, T. L., Lee, K. L., Kong, Y., Blumenthal, J. A., and Whalen, R. E. Type A behavior, hostility, and coronary atherosclerosis. *Psychosomatic Medicine*, 1980, in press.

Williams, R. B., Richardson, J. S., and Eichelman, B. S. Location of CNS neurons mediating the blood pressure fall after shock-induced fighting in the rat. *Journal of Behavioral Medicine*, 1978, *1*, 177–185.

Chapter 2

Physiologic and Neurobiologic Mechanisms in Arteriosclerosis

Floyd E. Bloom and J. Alan Herd

Substantial bodies of emerging evidence indicate that hormonal functions are regulated by the central nervous system. For example, the neuronal and hormonal information generated by the pancreatic islets and other metabolic units that contribute to nutrient homeostasis are monitored and regulated by higher brain centers. New methods for tracing neuroanatomical circuits have been able to specify the route taken within the peripheral and central nervous systems by visceral afferent fibers and the manner by which those primary visceral afferent target neurons are connected with other functionally defined levels of central integration. In some cases, though only a minority at best, these circuits can be attributed to specific neurotransmitters; in other major routes, in which visceral afferent and efferent information is also processed, the transmitter substances remain undefined. These circuits established for the rodent central nervous system have yet to be confirmed in primate brains or man, but general principles of comparative neuroanatomy suggest that major divergences will probably be in small details of microstructure. Many of the peripheral neuropeptides may also exert CNS selective effects, which can modify or antagonize the normal blood-born signals for insulin or glucagon secretion apart from direct actions on blood pressure, movement, or body temperature. The role of the CNS in the elicitation of diabetes-like imbalances of glucose and hormone levels can also be a direct result of central manipulations, and specifically the syndrome of stress hyperglycemia can acutely simulate the physiological changes seen in diabetes mellitus of adult onset.

A general hypothesis to be tested has emerged from physiologic studies as well as from behavioral studies designed to determine the manner by which

Floyd E. Bloom • The Salk Institute, P. O. Box 85800, San Diego, California 92138. J. Alan Herd • Sid. W. Richardson Institute for Preventive Medicine, The Methodist Hospital, 6565 Fannin, MS S400, Houston, Texas 77030.

appetitive acts are initiated and modulated. This hypothesis centers on the concept of Set Points—critical values of determining physiological variables (blood sugar, body weight, body temperature) to which different systems contribute in dynamic regulatory actions. Similarly, as concepts of the control of appetitive ingestion and its satiety control have been extended, it has become clear that the concept of opposing centers of eating is oversimplified, and that a set point for ingestion or satiety may be subject to a whole range of both motivational and physiological factors that could contribute to obesity. Recent evidence also indicates that there are hormonal receptors in the CNS and that these receptors have distinctive neuroanatomical distributions.

Mechanisms whereby behavioral processes might influence neuroendocrine activity have been demonstrated through studies of neuropeptides. This new family of neurochemicals includes many substances first identified as hormones secreted by the pituitary gland, such as ACTH, or as hormones secreted by the hypothalamus, such as thyrotropin-releasing factor, the molecule by which the hypothalamus regulates through the pituitary the functions of the thyroid gland (Guillemin, Yamazaki, Jutisz, and Sakiz, 1962).

The neuropeptides most identified with behavioral processes are the enkephalins and the endorphins. These neuropeptides appear to regulate neural and endocrine functions that play a role in behavioral adjustments to physical and psychological stimuli. The discovery of these peptides arose from the demonstrations that radioactively-labeled opiate compounds bind to opiate receptors concentrated in areas of the brain involved in the perception of pain and in integration of emotional behavior. Results of further studies indicate that endorphins are released from the pituitary gland in response to physical and psychological stimuli that cause secretion of ACTH. The relation between ACTH and endorphins has been shown by neurochemical studies.

Mains, Eipper, and Lin (1977) have shown that ACTH and beta-endorphin are products of a much larger precursor glycoprotein molecule. This molecule is present in cells of the anterior and intermediate lobes of the pituitary gland (Moon, Li, and Jennings, 1973; Pelletier, Lecleve, Labrie, Cote, Chretien, and Lis, 1977), and there is immunocytochemical evidence that endorphins and ACTH coexist within the brain (Krieger and Liotta, 1979). Stimuli known to activate the pituitary–adrenal system have been reported to decrease anterior pituitary endorphin content and to produce a parallel increase in plasma endorphin and ACTH levels (Guillemin, Vargo, Rossier, Minick, Ling, Rivier, Vale, and Bloom, 1977; Baizman, Cox, Osman, and Goldstein, 1979; Wiedemann, Satio, Linfoot, and Li, 1979). Alterations in brain levels have been reported in animals exposed to similar experimental procedures (Donovan, 1978). However, other studies have demonstrated that increased plasma levels of beta-endorphin may not be accompanied by increases in concentrations of beta-endorphin in the brain (Rossier, French, Rivier, Ling, Guillemin and Bloom, 1977).

The fact that endorphin is released from the pituitary along with ACTH suggests that it may function as a trophic hormone at peripheral organs. For example, morphine has been shown to stimulate release of epinephrine from the adrenal medulla even after complete denervation of the adrenal gland (Anderson and Slotkin, 1976). On the basis of evidence available, Axelrod (1972) has described a hypothalamic–anterior pituitary system that modifies epinephrine release from the adrenal medulla.

In the brain, endorphins apparently function to modulate neural systems that play a role in behavioral response to physical and psychological stimuli. For example, stimuli that produce an increase in blood or brain levels of endorphins also produce analgesia (Spiaggia, Bodnar, Kelly, and Glusman, 1979). However, administration of naloxone hydrochloride, a specific narcotic receptor antagonist, may not block this form of analgesia suggesting that nonopiate pathways may be involved as well (Grevert and Goldstein, 1978). In addition, results of other experiments (Fanselow, 1979) demonstrate that administration of naloxone has behavior effects consistent with the hypothesis that painful stimuli cause release of endorphins within the brain which attenuate the physiological and behavioral response to those stimuli.

The concept underlying the neuroendocrine correlates of behavioral processes arose with Cannon (1936) and was elaborated by Selye (1956). The original concept was one of a general physiological response, including adrenal medullary, adrenal cortical, and sympathetic nervous system responses to physical and psychological stimuli. Refinements of the original concept have included identification of different physiological and behavioral responses to acute and chronic conditions and different responses to anticipated vs. current stimulation. Mason (1968, 1964), in reviewing the results of neuroendocrine research on the pituitary–adrenal–cortical system, concluded that most studies of acute anticipated stimulations showed an increase in cortisol levels. Selye had proposed that chronic conditions of stimulation may lead to a decrease in adrenal cortical secretion by exhausting adrenal output capacity. However, physiological and behavioral responses to chronic conditions are less predictable than responses to acute stimulation (Mason, 1964; Bourne, Rose and Mason, 1967; Friedman, Mason and Hamburg, 1963).

The association between neuroendocrine processes and arteriosclerosis is evident in several clinical and experimental studies. Troxler, Sprague, Albanese, and Thompson (1977) found a significant correlation between elevated serial morning plasma cortisol levels and moderate to severe coronary arteriosclerosis. This association was noted in asymptomatic male subjects who had coronary angiography as part of their evaluation at the USAF School of

Aerospace Medicine. In these men plasma cortisol was second only to serum cholesterol as a discriminator between coronary disease and nondiseased individuals. Rheumatoid arthritis patients treated with corticosteroids are reported to have a threefold increase in arteriosclerosis (Kalbak, 1972), and corticosteroids have been reported to accelerate coronary atherosclerosis in patients with lupus erythematosus (Bulkley and Roberts, 1975). In addition, Friedman, Byers, and Rosenman (1972) have reported findings of higher levels of ACTH in individuals with Type-A behavior patterns than in individuals with Type-B patterns (Friedman, Byers, and Rosenman, 1972). Those individuals with Type-A behavior patterns who display impatience and an intense competitive drive coupled with hostility have been shown in prospective studies to have about twice the rate of new coronary heart disease as men lacking those characteristics (Rosenman, Friedman, Strauss, Jenkins, Zyanski, and Wurm, 1970; Rosenman, Brand, Jenkins, Friedman, Strauss, and Wurm, 1975; Rosenman, Brand, Jenkins, Friedman, Strauss and Wurm, 1975).

The influence of cortisol on development of atherosclerosis was tested experimentally in cynomolgus monkeys. Sprague, Troxler, Peterson, Schmidt, and Young (1980) administered cortisol orally to monkeys each day in doses which significantly diminished the diurnal variations of serum cholesterol without elevating daily peak cortisol concentrations. Animals which were fed a high cholesterol diet and received cortisol daily had a significantly greater involvement of aortic intimal surface area with atherosclerotic lesions compared to animals receiving a high cholesterol diet only. This atherogenic effect occurred independently of any effect of cortisol or serum or lipoprotein cholesterol concentrations. Other clinical studies indicated that lipid and related hormonal differences exist between individuals with Type-A behavior patterns and individuals with Type-B patterns (Rosenman and Friedman, 1974). Other research shows increased urinary norepinephrine excretions during the working day and elevated plasma norepinephrine responses to competition among men with Type-A behavior patterns. Greater elevations in plasma epinephrine were reported among individuals with Type-A behavior than among those with Type-B patterns during a competitive challenge in the laboratory (Glass, Krakoff, Contrada, Hilton, Kehoe, Mannucci, Collins, Snow, and Elting, 1979). The suggestion has been made that the elevated catecholamine responses may precipitate complications of coronary artery disease, such as acute myocardial infarction and sudden death.

The effects of social, psychological, and behavioral factors also can be demonstrated using measurements of regional cerebral blood flow and regional cerebral functional activity. Using noninvasive techniques, these measurements now can be carried out in humans during detailed psychological assessments. Studies using radioactive xenon to measure regional cerebral blood flow have revealed that blood flow through the left occipital and temporal-occipital sites was reduced as recognition accuracy increased (Wood, Taylor, Penny, and Stump, 1980). These results have been interpreted to imply that recognition memory tasks demand more blood flow in the distribution of the posterior cerebral arteries to the medial temporal lobes. Thus, these studies make it possible to determine which regions of the brain are involved in various behavioral processes and to observe how disease of blood vessels in various regions of the brain alters function of the brain.

A related radioactive technique, positron emission tomography (PET), also shows promise in measurements of regional cerebral blood flow and regional cerebral functional activity. Using an approach similar to that used in computer-assisted tomography scanning, the PET technique provides far greater resolution than conventional X-ray or radioisotope techniques. When radioactive nuclei such as ^{15}O, ^{13}N, ^{11}C, and ^{18}F emit a positron, those positrons encountering electrons in tissue annihilate each other and yield geometrically opposed photons that may be detected by gamma-ray detectors coupled with a coincidence counting circuit located at precisely opposite positions outside the skull. Using reconstruction tomography, a three-dimensional image of nuclear events is obtained (Eichling, Higgins, and Ter-Pogossian, 1977). Thus it should be possible to obtain a three-dimensional picture of neurochemical processes in any part of the brain under various psychological conditions. The PET method permits measurements of blood flow in deep structures of the brain (Phelps, Hoffman, Coleman, Welch, Raichle, Weiss, Sobel, and Ter-Pogossian, 1976), and metabolic processes can be measured with the use of ^{11}C-labeled glucose or ^{11}F-deoxyglucose (Reivich, Kuhl, Wolf, Greenberg, Phelps, Ido, Casella, Fowler, Gallagher, Hoffman, Alavi, and Sokoloff, 1977). Applications for PET in the future will include studies of neurotransmitter metabolism in association with various psychological, behavioral, and neuroendocrine processes.

Although the neurobiologic links between behavioral processes and arteriosclerosis are still unknown, a general concept is emerging. Physiological and psychological stimuli that are aversive modify both adrenal medullary and adrenal cortical activity through secretion of neuropeptides. These neuropeptides also modify behavior and may be conditioned by environmental stimuli to be secreted even in the absence of painful stimuli. Adrenal medullary and adrenal cortical secretions appear to influence the development of arteriosclerosis and precipitation of complications. However, the pathophysiological mechanisms influencing arteriosclerosis under these conditions are still unknown.

Psychologic stress research in humans has developed a large body of evidence indicating that the anticipation of threat to the integrity of the individual may lead to important changes, not only in thought, feeling, and action, but also in endocrine and autonomic processes, and hence in a wide variety of visceral functions. Much work in this field has centered on the changes in adrenocortical functioning that occur in association with emotional stress. Investigators have generally found the adrenal cortex to be stimulated via the brain under environmental conditions perceived by a person as threatening. In some studies, it has been possible systematically to correlate the extent of emotional distress with plasma and urinary corticosteroids assessed independently. Similar studies relying upon biochemical methods for measurement of epinephrine, norepinephrine, and aldosterone under conditions of emotional distress have yielded similar results. Moreover, experimental studies of monkeys and humans have generated substantial data indicating a linkage between emotional responses and adrenal function (Hamburg, 1967). Overall, the evidence clearly indicates that emotional distress in humans is associated with elevated blood levels and urinary excretion of adrenal hormones.

The most reasonable interpretation of these elevations is that they reflect increased secretion of the principal hormones of the adrenal gland, involving both the cortex and the medulla. These increases are associated with a wide variety of stressful conditions and emotional responses—not only with stresses that directly threaten the physical survival of the person but also with stresses that threaten self-respect or crucial human relationships. These responses appear to be part of a complex set of metabolic and cardiovascular adjustments in anticipation of vigorous action, a vital feature of adaptation over the long course of human evolution.

A number of studies have been undertaken in attempts to identify psychophysiological and neuroendocrine mechanisms that might account for the increased CHD rates observed among Type-A persons (Glass, Krakoff, Contrada, Hilton, Mannucci, Collins, Snow, and Elting, 1980). When challenged to perform a variety of behavioral tasks, Type As are observed to show greater increases in heart rate, blood pressure, and catecholamine secretion. Among those psychological characteristics which have been implicated in the tendency to display overt Type-A behavior under challenge are increased need for control (Glass, 1977), increased self-involvement (Scherwitz, Berton, and Leventhal, 1978), and increased levels of hostility (Matthews, Glass, Rosenman, and Bortner, 1977).

The physiological responses to physical and psychological stressors are similar in that several neurogenic, endocrine, and metabolic effects are elicited by both types of stressors. The neurogenic effects include alterations in central integrative mechanisms, an increase in sympathetic nervous system activity, and a decrease in parasympathetic activity. The endocrine effects include increased rates of secretion and increased circulating levels of epinephrine, norepinephrine, cortisol, growth hormone, plasma renin activity, and angiotensin II.

The linkage between neurobiologic and arteriosclerotic processes has been most extensively studied in terms of the neuroendocrine plasma–lipid relationship. Catecholamines are probably the most important factors promoting lipid mobilization (Heindel, Orci, and Jeanrenaud, 1975). They are effective in mobilizing lipid from adipose tissue (Rosell and Belfrage, 1975) both by their liberation from noradrenergic nerve terminals in adipose tissue and by their secretion from the adrenal medulla into the blood and arrival in adipose tissue through the circulation (Heindel *et al.*, 1975).

When triglyceride stores in adipose tissue are mobilized, they are hydrolyzed to free fatty acids (FFA) and glycerol (Heindel *et al.*, 1975). These FFA are utilized by skeletal muscle and myocardium in the production of energy (Zierler, Maseri, Klassen, Rabinowitz, and Burgess, 1968). They are both stored in these tissues as triglycerides and used directly. The FFA not stored or utilized in the production of energy are eventually taken up by the adipose tissues or by the liver (Shapiro, 1965). Those taken up by the liver are formed into triglycerides and secreted as a component of very low density lipoproteins (VLDL), (Schonfeld and Pfleger, 1971).

The effects of neuroendocrine factors on cardiovascular diseases involve many components and mechanisms leading to arteriosclerosis, including damage to vascular endothelium and proliferation of vascular smooth muscle cells (Ross and Glomset, 1976). Damage to vascular endothelium may be the result of hemodynamic factors such as high levels of arterial blood pressure or turbulence at bifurcations in large arteries. It also may be the result of chemical factors such as high blood levels of very low density lipoproteins or other circulating substances (Ross and Harker, 1976). Proliferation of vascular smooth muscle also may be influenced by several different factors. Tissue cultures of vascular smooth muscle cells proliferate most readily in the presence of VLDL (Ross and Glomset, 1973). Proliferation *in vitro* also is enhanced by the presence of insulin in the culture medium (Stout, Bierman, and Ross, 1975). These results from laboratory studies of atherogenic factors suggest mechanisms whereby physiological processes associated with neuroendocrine and metabolic factors may influence atherogenesis in humans and experimental animals.

The rate at which the liver secretes VLDL is determined partly by the rate at which it synthesizes FFA from carbohydrates and partly by the rate at which it receives FFA in the blood. In the fasting state, the secretion of VLDL by the liver is determined principally by the levels of FFA in the blood (Basso and Havel, 1970). These levels, in turn, are determined principally by the effects of catecholamines on adipose tissue and rates of energy production.

Results obtained recently from several laboratories also suggest that FAA may play an important role in the pathogenesis of cardiovascular disease. FFA may predispose to development of atherosclerosis, and they may precipitate cardiac arrhythmias. Therefore, individuals who maintain high levels of FFA (Nestel, Ishikawa, and Goldrick, 1978) may be at high risk for developing cardiovascular disease.

FFA levels in plasma have at least two important physiological effects. Increases in FFA levels increase intensity of platelet aggregation (Burstein, Berns, Heldenberg, Kahn, Werbin, and Tamir, 1978). Since platelet aggregation plays a major role in evolution of atheroma, it influences the basic disease process. Increases in FFA levels also increase myocardial oxygen requirements (Simonsen and Kjekshus, 1978). Catecholamines sensitize the heart to FFA in such a way that FFA account for a major part of the increased myocardial oxygen requirement during SAM stimulation. In the presence of coronary artery disease, increases in FFA and increases in myocardial oxygen requirement or decreases in myocardial oxygen supply frequently cause serious ventricular arrhythmias.

The physiologic and neurobiologic mechanisms linking behavioral factors to arteriosclerosis are worthy of further research. The hypothesis is that behavioral factors elicit sympathetic adrenomedullary and pituitary adrenocortical activity which may cause arteriosclerosis, myocardial infarction, and ventricular arrhythmias. Although neuroendocrine function is essential for normal physiological function, elevations in neuroendocrine activity may predispose to arteriosclerosis and coronary heart disease.

References

Anderson, T. R., and Slotkin, T. A. Teh role of neural input in the effects of morphine on the rat adrenal medulla. *Biochemical Pharmacology*, 1976, *25*, 1071–1074.

Axelrod, J. Dopamine-beta-hydroxylase: Regulation of its synthesis and release from nerve terminals. *Pharmacology Review*, 1972, *24*, 233–243.

Baizman, E. R., Cox, B. M., Osman, O. H., and Goldstein, A. Experimental alterations of endorphin levels in rat pituitary. *Neuroendocrinology*, 1979, *28*, 402–424.

Basso, L. V., and Havel, R. J. Hepatic metabolism of free fatty acids in normal and diabetic dogs. *Journal of Clinical Investigation*, 1970, *49*, 537–547.

Bourne, P. G., Rose, R. M., and Mason, J. W. Urinary 17-OHCS levels. Data on seven helicopter ambulance medics in combat. *Archives of General Psychiatry*, 1967, *17*, 104–110.

Bulkley, B. H., and Roberts, W. C. The heart in systemic lupus erythematosis and the changes induced in it by corticosteroid therapy. *American Journal of Medicine*, 1975, *58*, 243–264.

Burstein, Y., Berns, L., Heldenberg, D., Kahn, Y., Werbin, B. Z., and Tamir, I. Increase in platelet aggregation following a rise in plasma free fatty acids. *American Journal of Hematology*, 1978, *4*, 17–22.

Cannon, W. B. The role of emotion in disease. *Annals of Internal Medicine*, 1936, *9*, 1453–1465.

Donovan, B. T. The behavioral actions of the hypothalamic peptides: a review. *Psychological Medicine*, 1978, *8*, 305–316.

Eichling, J. O., Higgins, C. S., and Ter-Pogossian, M. M. Determination of radionuclide concentrations with positron CT scanning (PETT): Concise communication. *Journal of Nuclear Medicine*, 1977, *18*, 845–847.

Fanselow, M. Naloxone attenuates rats preference for signaled shock. *Physiology and Psychology*, 1979, *7*, 70–74.

Friedman, M., Byers, S. O., and Rosenman, R. H. Plasma ACTH and cortisol concentration of coronary-prone subjects. *Proceedings of Society for Experimental Biological Medicine*, 1972, *140*, 681–684.

Friedman, S. B., Mason, J. W., and Hamburg, D. A. Urinary 17-hydroxycorticosteroid levels in parents of children with neoplastic disease: A study of chronic psychological stress. *Psychosomatic Medicine*, 1963, *25*, 364–376.

Glass, D. C. *Behavior patterns, stress, and coronary disease*. Hillsdale, New Jersey: Lawrence Erlbaum Associates, 1977.

Glass, D. C., Krakoff, L. R., Contrada, R., Hilton, W. F., Mannucci, E. G., Collins, C., Snow, B., and Elting, E. Effect of harassment and competition upon cardiovascular and plasma catecholamine responses in Type A and Type B individuals. *Psychophysiology*, 1980, *17*, 453–463.

Glass, D. C., Krakoff, L. R., Contrada, R., Hilton, W. F., Kehoe, K., Mannucci, E. G., Collins, C., Snow, B., and Elting, E. Effect of harassment and competition upon cardiovascular and catecholaminic responses in Type A and Type B individuals. California Panel on Biobehavior to the Working Group on Arteriosclerosis, La Jolla, California. Washington: Government Printing Office.

Grevert, P., and Goldstein, A. Endorphins: Naloxone fails to alter experimental pain or mood in humans. *Science*, 1978, *199*, 1093–1095.

Guillemin, R., Vargo, T., Rossier, J., Minick, S., Ling, N., Rivier, C., Vale, W., and Bloom, F. β-Endorphin and adrenocorticotropin are secreted concomitantly by the pituitary gland. *Science*, 1977, *197*, 1367–1369.

Guillemin, R., Yamazaki, E., Jutisz, M., and Sakiz, E. Presence dans un extrait de tissus hypothalaminques d'une substance stimulant la secretion de l'hormone hypophysaire thyreotrope (TSH). Premiere purification par filtration sur gel Sephadex. *Compte Rendus Hebdomadaires des Seances de L'Academie des Sciences. D: Sciences Naturelles (Paris)*, 1962, *255*, 1018–1020.

Hamburg, D. Genetics of adrenocortical hormone metabolism in relation to psychological stress. In J. Hirsch (Ed.), *Behavior-genetic analysis*, New York: McGraw-Hill, 1967.

Heindel, J. J., Orci, L., and Jeanrenaud, B. Fat mobilization and its regulation by hormones and drugs in white adipose tissue. In Masoro, E. J. (Ed.), *International Encyclopedia of Pharmacology and Therapeutics*. Pharmacology of lipid transport and atherosclerotic processes. Oxford: Pergamon Press, 1975.

Kalbak, K. Incidence of arteriosclerosis in patients with rheumatoid arthritis receiving long-term corticosteroid therapy. *Annals of Rheumatic Disorders*, 1972, *31*, 196–200.

Krieger, D. T., and Liotta, A. S. Pituitary hormones in brain: Where, how, and why. *Science*, 1979, *205*, 366–372.

Mains, R. E., Eipper, B. A., and Lin, N. Common precursor to corticotropins and endorphins. *Proceedings of the National Academy of Science, USA*, 1977, *74*, 3014–3018.

Mason, J. W. Psychoendocrine approaches in stress research. In *Symposium on Medical Aspects of Stress in the Military Climate*. Walter Reed Army Institute of Research, Washington, D.C., 1964.

Mason, J. W. A review of psychoendocrine research on the pituitary–adrenal cortical system. *Psychosomatic Medicine*, 1968, *30*, 576–607.

Matthews, K. A., Glass, D. C., Rosenman, R. H., and Bortner, R. W. Competitive drive, pattern A, and coronary heart disease: A further analysis of some data from the Western Collaborative Group Study. *Journal of Chronic Diseases*, 1977, *30*, 489–498.

Moon, H. D., Li, C. H., and Jennings, B. M. Immunohistochemical and histochemical studies of pituitary β-lipotrophs. *Anatomical Record*, 1973, *175*, 529–538.

Nestel, P. J., Ishikawa, T., and Goldrick, R. B. Diminished plasma-free fatty acid clearance in obese subjects. *Metabolism*, 1978, *27*, 589–597.

Pelletier, G., Lecleve, R., Labrie, F., Cote, J., Chretien, M., and Lis, M. Immunohistochemical localization of β-lipotropic hormone in the pituitary gland. *Endocrinology*, 1977, *100*, 770–776.

Phelps, M. E., Hoffman, E. J., Coleman, R. E., Welch, M. J., Raichle, M. E., Weiss, E. S., Sobel, B. E., and Ter-Pogossian, M. M. Tomographic images of blood pool and perfusion in brain and heart. *Journal of Nuclear Medicine*, 1976, *17*, 603–612.

Reivich, M., Kuhl, D., Wolf, A., Greenberg, J., Phelps, M., Ido, T., Casella, V., Fowler, J., Gallagher, B., Hoffman, E., Alvi, A., and Sokoloff, L. Measurement of local cerebral glucose metabolism in man with ^{18}F-2-fluorodeoxy-D-glucose. *Acta Neurologica Scandinavica Supplementum*, 1977, *56*, 190–191.

Rosell, S., and Belfrage, E. Adrenergic receptors in adipose tissue and their relation to adrenergic innervation. *Nature*, 1975, *253*, 738.

Rosenman, R., Brand, R. J., Jenkins, C. D., Friedman, M., Straus, R., and Wurm, M. Coronary heart disease in the Western Collaborative Group Study. *Journal of the American Medical Association*, 1975, *233*, 872–877.

Rosenman, R. H., Brand, R. J., Jenkins, C. D., Friedman, M., Straus, R. and Wurm, M. Coronary heart disease in the Western Collaborative Group Study: Final follow-up experience of 8½ years. *Journal of the American Medical Association*, 1975, *223*, 872–877.

Rosenman, R. H., and Friedman, M. Neurogenic factors in pathogenesis of coronary heart disease. *Medical Clinics of North America*, 1974, *58*, 269–279.

Rosenman, R. H., Friedman, M., Straus, R., Jenkins, C. D., Zyanski, S. J., and Wurm, M. Coronary heart disease in the Western Collaborative Group Study. *Journal of Chronic Diseases*, 1970, *23*, 173–190.

Ross, R., and Glomset, J. A. Atherosclerosis and the arterial smooth muscle cell. *Science*, 1973, *180*, 1332–1339.

Ross, R., and Glomset, J. A. The pathogenesis of atherosclerosis. *New England Journal of Medicine*, 1976, *295*, 369–377.

Ross, R. and Harker, L. Hyperlipidemia and atherosclerosis. *Science*, 1976, *193*, 1094–1100.

Rossier, J., French, E. D., Rivier, C., Ling, N., Guillemin, R., and Bloom, F. E. Foot-shock induced stress increases β-endorphin levels in blood but not brain. *Nature*, 1977, *270*, 618–620.

Scherwitz, L., Berton, K., and Leventhal, H. Type A behavior, self-involvement, and cardiovascular response. *Psychosomatic Medicine*, 1978, *XL*, 593–609.

Schonfeld, G., and Pfleger, B. Utilization of exogenous free fatty acids for the production of very low density lipoprotein triglyceride by livers of carbohydrate-fed rats. *Journal of Lipid Research*, 1971, *12*, 614–621.

Selye, H. *The Stress of Life*. New York: McGraw-Hill, 1956.

Shapiro, B. Triglyceride metabolism. In A. E. Renold and G. F. Cahill, Jr. (Eds.), *Handbook of Physiology*, Section 5, Adipose Tissue. Washington, D.C.: American Physiology Society, 1965.

Simonsen, S. and Kjekshus, J. K. The effect of free fatty acids on myocardial oxygen consumption during a trial pacing and catecholamine infusion in man. *Circulation*, 1978, *58*, 484–491.

Spiaggia, A., Bodnar, R., Kelly, D., and Glusman, M. Opiate and nonopiate mechanisms of stress-induced analgesia: Cross-tolerance between stressors. *Pharmacology, Biochemistry and Behavavior*, 1979, *10*, 761–766.

Sprague, E. A., Troxler, R. G., Peterson, D. F., Schmidt, R. E., and Young, J. T. Effect of cortisol on the development of atherosclerosis in cynomolgus monkeys. In S. S. Kalter (Ed.), *The use of nonhuman primates in cardiovascular diseases*. Austin: University of Texas Press, 1980.

Stout, R. W., Bierman, E. L. and Ross, R. Effect of insulin on the proliferation of cultured primate arterial smooth muscle cells. *Circulation Research*, 1975, *36*, 319–327.

Troxler, R. G., Sprague, E. A., Albanese, R. A. and Thompson, A. J. The association of elevated plasma cortisol and early atherosclerosis as demonstrated by coronary angiography. *Atherosclerosis*, 1977, *26*, 151–162.

Wiedemann, E., Satio, T., Linfoot, J. A., and Li, C. H. Specific radioimmunoassay of human beta-endorphin in unextracted plasma. *Journal of Clinical Endocrinology and Metabolism*, 1979, *49*, 478–480.

Wood, R., Taylor, M. E., Penny, R., and Stump, D. Regional cerebral blood flow response to recognition memory versus semantic classification tasks. *Brain Language*, 1980, *9*, 113–122.

Zierler, K. L., Maseri, A., Klassen, D., Rabinowitz, D., and Burgess, J. Muscle metabolism during exercise in man. *Transactions of the Association of American Physicians*, 1968, *81*, 266–273.

Chapter 3

Psychosocial Influences and the Pathogenesis of Arteriosclerosis[1]

David C. Glass

A large body of data on the role of behavioral factors in the etiology and pathogenesis of coronary heart disease (CHD) has accumulated over the past few decades (Jenkins, 1976). Two promising variables have been identified: psychological stress and what has been called the Type-A coronary-prone behavior pattern (Rosenman and Friedman, 1974). Stress may be defined as an internal state of the individual when he is faced with threats to his physical and/or psychic well-being. Following Cox (1978) and Lazarus (1966), the perception of threat is believed to arise from a comparison between the demands on the individual and his assessment of his ability to cope. A perceived imbalance in this mechanism gives rise to the experience of stress and to the stress response, which may be behavioral and/or physiologic in nature.

An individual who shows Type-A behavior is competitive and hard-driving, time-urgent and impatient, hostile and aggressive. By contrast, Type-B individuals are noted for the relative absence of these characteristics. Type A behavior is the outcome of a person–situation interaction. It is elicited only in the presence of appropriate environmental circumstances, including the challenge of doing well at a difficult task, or the stress of uncontrollable aversive stimulation (Glass, 1977). Type-A behavior is, moreover, one extreme of a bipolar continuum; the other extreme is defined by Type-B behavior.

1. Psychological Stress and CHD

Several classes of psychological stressors have been linked to the major cardiovascular disorders, including dissatisfaction with marital relationships

David C. Glass • Graduate School and University Center, City University of New York, 33 West 42 Street, New York, New York 10036.

and other interpersonal relations. Excessive work and responsibility, which leads to feelings that job demands are beyond the person's control, has been implicated, as well, in the development of coronary disease. Additionally, there are data to suggest that acutely stressful events, such as the death of a close relative or the sudden loss of self-esteem, increase the likelihood of a coronary event. More detailed discussion of these matters can be found in a number of recent review papers and books (Jenkins, 1976; Glass, 1977).

The physiological mechanisms whereby psychological stress may enhance the development of cardiac disorders include repeated increases in serum lipids (e.g., cholesterol) and blood pressure; acceleration of the rate of damage to the coronary arteries over time; facilitation of platelet aggregation; induction of myocardial lesions; and precipitation of cardiac arrhythmias. It is believed in some quarters that these effects are mediated by enhanced activity of the sympathetic nervous system and consequent discharge of catecholamines such as epinephrine and norepinephrine (Haft, 1974; Eliot, 1979).

2. Type-A Behavior and CHD

Perhaps the most thoroughly studied behavioral factor contributing to coronary disease is the Type-A behavior pattern. Issues of Type-A measurement and classification cannot be discussed within the confines of this paper. Suffice it to note that the major diagnostic tools are a structured interview developed by Friedman and Rosenman, and a self-administered questionnaire called the "Jenkins Activity Survey for Health Prediction." More detailed considerations of measurement can be found in Dembroski, Weiss, Shields, Haynes, and Feinlieb (1978a).

The strongest available evidence for the association between Type-A behavior and CHD comes from the Western Collaborative Group Study (WCGS). The results indicate that Type-A men experienced about twice the incidence of acute clinical events over an 8½ year follow-up period compared to Type-B men (Rosenman, Brand, Jenkins, Friedman, Straus, and Wurm, 1975). This difference occurred independently of other risk factors, including total serum cholesterol, systolic blood pressure, and daily cigarette smoking. Although the evidence is not unequivocal (Dimsdale, Hackett, Hutter, and Block, 1980), other research using coronary arteriography has documented more severe occlusion of the coronary arteries in Type-A compared to Type-B patients (Zyzanski, Jenkins, Ryan, Flessas, and Everist, 1976; Blumenthal, Williams, Kong, Schanberg, and Thompson, 1978; Frank, Heller, Kornfield, Sporn, and Weiss, 1978).

If we accept the results showing a linkage between Type A and coronary disease, the next question concerns the physiological mechanisms underlying the association. Clinical studies indicate that lipid and related hormonal differences exist between Type-A and Type-B individuals (Rosenman and Friedman, 1974). Other research shows increased urinary norepinephrine (NE) excretion in Type-A individuals during an active working day compared to during more sedentary evening activities (Byers, Friedman, Rosenman, and Freed, 1962). Also relevant here is a study indicating elevated plasma NE responses to competition and stress among Type-A compared to Type-B men (Friedman, Byers, Diamant, and Rosenman, 1975).

Elevated catecholamine responses are likely to be associated with changes in cardiovascular function, which could be crucial in the potentiation of CHD and sudden death. A number of studies have, in fact, shown that As display greater episodic increases in systolic blood pressure and heart rate than do Bs in stressful and challenging situations (Dembroski, Weiss, Shields, Haynes, and Feinlieb, 1978a; Manuck and Schaefer, 1978; Pittner and Houston, 1980). There is typically little difference between the two types of individuals in basal levels of these cardiovascular variables. The observation of heightened physiologic reactivity among As takes on added significance when viewed in the light of recent work by Manuck and Schaefer (1978), and Obrist, Grignolo, Hastrup, Koepke, Langer, Light, McCubbin, and Pollack (in press). These investigators have shown that cardiovascular reactivity is stable over time, and that it is related to parental history of high blood pressure—a risk factor for hypertension which, in turn, enhances the likelihood of CHD.

Research in my laboratory has also documented cardiovascular and plasma catecholamine reactivity in As compared to Bs, when both types of subjects are exposed to stressful competition. Subjects were Type-A and B men from the work force of the New York City Transit authority (Glass, Krakoff, Contrada, Hilton, Kehoe, Mannucci, Collins, Snow, and Elting, 1980). All were free of the major risk factors for CHD, including hypertension, diabetes mellitus, excessive cigarette smoking, and elevated levels of serum cholesterol. The purpose of the study, conducted in collaboration with Dr. Lawrence R. Krakoff of the Mount Sinai Medical School in New York, was to assess the effects upon arterial pressure, heart rate, and plasma catecholamines of simple competition in a game *vs.* competition with a hostile and harassing opponent (experimental confederate).

Arterial pressure (SBP and DBP) and heart rate (HR) were monitored every two minutes over the course of the session. Blood pressure determinations were made using a Roche Arteriosonde, which ultrasonically detects arterial wall motion. Heart rate was measured with a photocell plethysmograph, which allowed the recording of digital pulsation. Blood for plasma epinephrine (E) and norepinephrine (NE) was obtained by means of an in-dwelling venous catheter. Samples were taken at the end of the baseline period, and again after the third and sixth games. A fourth sample was drawn after the ninth and final game, when the subject had lost to the opponent who received the gift certificate.

The results for each dependent measure were averaged for each of the three experimental periods (i.e., blocks of three-game segments) and compared to their respective baseline values. Statistical analyses of these data revealed the following effects:

1. There were no differences between As and Bs, or between the two experimental conditions, in mean baseline values for SBP, DBP, HR, plasma E, and plasma NE (p's > .20).
2. All increases in the five dependent measures relative to baseline values were statistically significant (p's < .05).
3. The increase in SBP for Harassed As was significantly greater than for No Harassed As and each of the two B groups (p's < .05), (Table 1).
4. The effect of harassment of DBP was near-significant at the .06 level, but there was no difference between As and Bs (p's > .20).
5. The increase in HR for the As was significantly greater than for the Bs (p < .05), and the source of this difference was attributed to the elevations of the Harassed As (Table 2).
6. The increase in plasma E for the Harassed As was also significantly greater than for any other experimental group (Table 3).
7. Although the mean change-scores for the plasma NE were in the same direction as those for plasma E, they did not attain acceptable levels of statistical significance due to within-groups variability.

In conclusion:

1. Competition elicits significant and similar increases in BP, HR, and plasma catecholamines in Type-A and -B men.
2. The effect of a hostile opponent causes no reliable differences in cardiovascular and plasma catecholamine responses in Type Bs.

Table 1
Systolic Blood Pressure: Mean Changes for Each Third of the Contest (in mmHg)

Condition	1st third	2nd third	Last third
Type A, harass	+37.5	+41.9	+38.5
Type A, no harass	+24.5	+27.9	+27.4
Type B, harass	+24.2	+26.5	+26.4
Type B, no harass	+24.1	+25.8	+25.3

Table 2
Heart Rate: Mean Changes for Each Third of the Contest (in bpm)

Condition	1st third	2nd third	Last third
Type A, harass	+26.1*	+24.6	+21.4
Type A, no harass	+15.6*	+17.8	+14.9
Type B, harass	+12.4	+13.4	+12.8
Type B, no harass	+14.0	+12.1	+10.9

*Apparatus failure interferred with the recording of heart rate from one case in this group.

Table 3
Plasma Epinephrine (E): Mean Changes during and after the Contest (in pg/ml)

Condition	During the contest	After the contest
Type A, harass	+100.2*	+120.5**
Type A, no harass	+ 3.7	+ 19.8
Type B, harass	+ 16.9*	+ 2.0**
Type B, no harass	+ 16.1	+ 20.0**

*Two cases each were lost from the Harass A and Harass B groups because of technical difficulties connected with assay and blood sampling procedures.
**Four cases (one Harass A, one Harass B, and two No Harass Bs) were eliminated from analysis of E levels in the third (i.e., "after-contest") blood sample. Hemolyzed samples made it virtually impossible to calculate accurate values for these subjects.

3. In As, however, the harassing opponent elicits greater increases in SBP, HR, and plasma E during competition.
4. It would appear that behavior pattern A is predisposed selectively to enhanced reaction to hostile interactions, but competition alone does not distinguish between As and Bs.

This experiment, as well as others conducted in my laboratory over the past year, suggests some of the cardiovascular and neuroendocrine variables that might account for the greater tendency of Type-A individuals to develop coronary disease. It is speculative to argue that excess production of epinephrine and heightened systolic pressor responses serve as the intermediary process by which Type-A behavior enhances the risk of cardiovascular disease. However, it is probably reasonable to assume that the observed physiologic responses of Type As to environmental stress are mediated via the sympathetic nervous system.

3. Type-A Behavior Pattern as a Psychological Construct

Thus far, this discussion has emphasized physiological mechanisms underlying the Type A–coronary disease relationship. There is a more basic issue, namely, whether behavior pattern A is a valid psychological construct.

A large proportion of the population is typically classified as Type A, estimates having ranged from 45% to as high as 76% in some populations (Rosenman, Friedman, Straus, Wurm, Kositchek, Hahn, and Werthessen, 1964; Howard, Cunningham, and Rechnitzer, 1976). Nevertheless, there is a relatively low incidence of CHD among As, albeit significantly greater than in Bs (Rosenman et al., 1975). Therefore, the causal mechanisms underlying cardiovascular disease may not be distributed evenly throughout the Type-A group. It is possible that some facets of behavior pattern A have little or no relationship to the disease, as they appear in all As rather than only in those who show increased risk. On the other hand, it may be that Type-A behaviors of any kind always lead to physiologic changes culminating in illness, but CHD occurs only in As who lack psychological and/or physiological protective mechanisms. In either case, these considerations underscore the importance of understanding the psychological mechanisms that give rise to and sustain Type-A behavior.

To date there have been three major approaches to identifying such mechanisms. The first derives from a factor analysis of the structured interview responses of 186 men from the WCGS (Matthews, Glass, Rosenman, and Bortner, 1977). Although five primary factors were revealed, only two—competitive drive and impatience—were associated with the later onset of clinical CHD. Subsequent analyses indicated that of the more than forty interview ratings, only seven items discriminated CHD cases from age-matched healthy controls. Of the seven, four items related directly to hostility, one was concerned with competitiveness, and the remaining two dealt with vigorousness of voice stylistics. Dembroski, MacDougall, Shields, Petitto, and Lushene (1978b) have developed a component scoring system for the structured interview based on the Matthews et al. findings (1977). The same dimensions that predicted CHD were found to predict experimentally-induced elevations in systolic blood pressure and heart rate.

A somewhat different approach to the association between Type-A behavior and CHD comes from the work of Scherwitz, Berton, and Leventhal (1978a). They identified and measured certain speech characteristics that occurred continuously in the structured interview. These characteristics were correlated with simultaneously occurring changes in heart rate, finger pulse amplitude, and blood pressure. Type-A individuals who used many self-references (I, me, my, mine) in answering the interview questions showed the highest levels of systolic blood pressure. By contrast, the Type-B group had very few significant correlates of self-references. These results suggest that self-involvement might account for both the speech characteristics and autonomic reactions of Type-A subjects. Indeed, there is evidence that individuals with acute self-awareness behave like Type As. For example, individuals whose attention is focused on themselves are likely to be aggressive when provoked (Scheier, 1976). While performing a task, self-aware individuals are more likely to compare their performance to their internal standards of excellence (Carver, Blaney, and Scheier, in press). To the extent that these standards are high, salient discrepancies between performance and goals may lead to excessive striving, frustration, and helplessness. Thus, Scherwitz *et al.* (1978a) suggest that the construct of self-involvement is ·useful not only because it may explain why Type-A behaviors arise, but also because its correlations with cardiovascular and behavioral variables underscore its potential importance as a key construct in explaining the linkage between Type-A behavior and CHD.

A third approach to the issue of mechanism comes from my own past work on Type-A behavior (Glass, 1977). In several studies, my students and I have found that Type As work hard to succeed, suppress subjective states (such as fatigue) that might interfere with task performance, conduct their activities at a rapid pace, and express hostility after being frustrated or harassed in their efforts at task completion. It might be argued that these behaviors reflect an attempt by the Type-A individual to assert and maintain conrol over stressful aspects of his environment. Type As engage in a continual struggle for control and, in consequence, appear hard-driving and aggressive, easily annoyed, and competitive. Furthermore, this struggle by As may lead them, when confronted by clear threats to that control, to increase their efforts to assert mastery. On the other hand, if these efforts meet with repeated failure, Type As might be expected to give up responding and act helpless. Stated somewhat differently, initial exposure to threatened loss of control accelerates control efforts on the part of As, whereas prolonged exposure leads to a *decrement* in these behaviors. This pattern of responding has been described elsewhere as *hyper-responsiveness* followed by *hypo-responsiveness* (Glass, 1977).

Experimental results with healthy human subjects reported elsewhere have tended to support the foregoing hypothesis (Glass, 1977). However, it should be emphasized that there is no evidence, to date, that bears on the interactive effects of depression and Type A on clinical CHD. Such an association would, of course, be consistent with the hypo-responsiveness part of our hypothesis. While it has been suggested that depression is a risk factor for a variety of illnesses and for delayed recovery from such illnesses (Greene, Moss, and Goldstein, 1974; Greene, Goldstein, and Moss, 1972; Engel, 1970), there are no data showing that depressed Type As are at greater risk for CHD than their Type-B counterparts. A test of this notion must await future research.

Central to our approach to understanding Type-A behavior is the idea that active efforts by Type As to control their environment are accompanied by sympathetic activation and elevated levels of circulating catecholamines. When efforts at control fail—as they inevitably will with an uncontrollable stressor—the theory predicts a shift from sympathetic to parasympathetic dominance. Such abrupt shifts have been implicated in sudden cardiac death (Engel, 1970). Still other research, some of which was cited earlier in this paper, indicates that elevated catecholamines are important factors in the pathogenesis of cardiovascular disease and acute clinical events.

These are three of many possible avenues for further differentiation of the psychological processes underlying Type-A behavior and the connection of these processes to physiological changes that may enhance cardiovascular risk. As these and other models are more fully developed and tested, we may expect the Type-A concept to be superceded by more differentiated factors that are more closely linked to pathophysiologic mechanisms (Scherwitz, Leventhal, Cleary, and Laman, 1978b).

4. Future Directions

Attention should be given to conceptualizing Type-A behavior and identifying the psychological processes that produce and sustain it. Most of the existing literature simply describes a behavior pattern and its behavioral and physiological correlates. We must define more precisely those behaviors in Type A that are risk-inducing. In this connection, serious consideration should be given to the notion that Type-A behavior is an outgrowth of person–situation interaction. It follows from this view that efforts need to be directed toward a delineation of the classes of environmental stimuli that elicit the primary facets of the behavior pattern. It is not enough to speak loosely about appropriately challenging and/or stressful events. We need to define these terms with precision. We need to specify the relevant parameters, such that we will be able to determine, a priori, which types and levels of stress or challenge are sufficient to produce Type-A behaviors and concomitant physiologic responses in both laboratory and field settings.

Another direction for future research should concentrate on linking the behavior pattern, or more appropriately the underlying dimensions of the pattern, to physiological processes believed to be routes to atherosclerosis and clinical CHD. Once such correlations are established, subsequent studies with animal models might be undertaken to elucidate cause-and-effect; that is, are the physiological changes observed in Type As under stress the result of behavioral responses or vice-versa? Indeed, both behavioral and physiological reactions may be consequences of a third, higher-order variable located in the brain. When such causal research is well underway, moderate-sized field studies aimed at evaluating whether the principles derived from psychophysiological experimentation can predict disease endpoints could be considered. At that time it might be appropriate to consider the advisability and feasibility of altering Type-A behavior, or at least those facets of the behavior pattern that have been established as enhancing the risk of cardiovascular disease.

Footnotes

1. A version of this paper appeared in J. Siegrist and M. J. Halhuber (Eds.), *Myocardial infarction and psychosocial risks.* New York: Springer-Verlag, 1980.

References

Blumenthal, J. A., Williams, R., Kong, Y., Schanberg, S. M., and Thompson, L. W. Type A behavior and angiographically documented coronary disease. *Circulation*, 1978, *58*, 634–639.

Byers, S. O., Friedman, M., Rosenman, R. H., and Freed, S. C. Excretion of VMA in men exhibiting behavior pattern (A) associated with high incidence of clinical coronary artery disease. *Federation Proceedings*, 1962, *21*, 99–101.

Carver, C. S., Blaney, P. H., and Scheier, M. F. Focus of attention, chronic expectancy, and responses to a feared stimulus. *Journal of Personality and Social Psychology*, 1979, *37*, 1186–1195.

Cox, T. *Stress*. Baltimore: University Park Press, 1978.

Dembroski, T. M., Weiss, S. M., Shields, J. L., Haynes, S. G., and Feinlieb, M. *Coronary-prone behavior.* New York: Springer-Verlag, 1978a.

Dembroski, T. M., MacDougall, J. M. Shields, J. L., Petitto, J., and Lushene, R. Components of the Type A coronary-prone behavior pattern and cardiovascular responses to psychomotor performance challenge. *Journal of Behavioral Medicine*, 1978b, *1*, 159–176.

Dimsdale, J. E., Hackett, T. P., Hutter, A. M., and Block, P. C. The risk of Type A-mediated coronary artery disease in different populations. *Psychosomatic Medicine*, 1980, *42*, 55–62.

Eliot, R. S. *Stress and the major cardiovascular disorders.* Mount Kisco, New York: Futura, 1979.

Engel, G. L. Sudden death and the 'medical model' in psychiatry. *Canadian Psychiatric Association Journal*, 1970, *15*, 527–538.

Frank, K. A., Heller, S. S., Kornfield, D. S., Sporn, A. A., and Weiss, M. B. Type A behavior pattern and coronary angiographic findings. *Journal of the American Medical Association*, 1978, *240*, 761–763.

Friedman, M., Byers, S. O., Diamant, J., and Rosenman, R. H. Plasma catecholamine response of coronary-prone subjects (Type A) to a specific challenge. *Metabolism*, 1975, *24*, 205–210.

Glass, D. C. *Behavior patterns, stress, and coronary disease.* Hillsdale, New Jersey: Lawrence Erlbaum Associates, 1977.

Glass, D. C., Krakoff, L. R., Contrada, R., Hilton, W. F., Kehoe, K., Mannucci, E. G., Collins, C., Snow, B., and Elting, E. Effect of harassment and competition upon cardiovascular and plasma catecholamine responses in Type A and Type B individuals. *Psychophysiology*, 1980, *17*, 453–463.

Greene, W. A., Moss, A. J., and Goldstein, S. Delay, denial and death in coronary heart disease. In R. S. Eliot (Ed.), *Stress and the heart*. Mount Kisco, New York: Futura, 1974.

Greene, W. A., Goldstein, S., and Moss, A. J. Psychosocial aspects of sudden death: A preliminary report. *Archives of Internal Medicine*, 1972, *129*, 725–731.

Haft, J. I. Cardiovascular injury induced by sympathetic catecholamines. *Progress in Cardiovascular Diseases*, 1974, *17*, 73–86.

Howard, J. H., Cunningham, D. A., and Rechnitzer, P. A. Health patterns associated with Type A behavior: A managerial population. *Journal of Human Stress*, 1976, *2*, 24–31.

Jenkins, C. D. Recent evidence supporting psychologic and social risk factors for coronary disease. *New England Journal of Medicine*, 1976, *294*, 987–994; 1033–1038.

Lazarus, R. S. *Psychological stress and the coping process*. New York: McGraw-Hill, 1966.

Manuck, S. B., and Schaefer, D. C. Stability of individual differences in cardiovascular reactivity. *Physiology and Behavior*, 1978, *21*, 675–678.

Matthews, K. A., Glass, D. C., Rosenman, R. H., and Bortner, R. W. Competitive drive, pattern A, and coronary heart disease: A further analysis of some data from the Western Collaborative Group Study. *Journal of Chronic Diseases*, 1977, *30*, 489–498.

Obrist, P. A., Grignolo, A., Hastrup, J. L., Koepke, J. P., Langer, A. W., Light, K. C., McCubbin, J. A., and Pollak, M. H. Behavioral-cardiac interactions in hypertension. In D. S. Krantz, A. Baum, and J. E. Singer (Eds.), *Handbook of psychology and health: Cardiovascular disorders*. Hillsdale, New Jersey: Lawrence Erlbaum Associates, in press.

Pittner, M. S., and Houston, B. K. Response to stress, cognitive coping strategies, and the Type A behavior pattern. *Journal of Personality and Social Psychology*, 1980, in press.

Rosenman, R. H., Friedman, M., Straus, R., Wurm, M., Kositchek, R., Hahn, W., and Werthessen, N. T. A predictive study of coronary heart disease: The Western Collaborative Group Study. *Journal of the American Medical Association*, 1964, *189*, 15–22.

Rosenman, R. H., and Friedman, M. Neurogenic factors in pathogenesis of coronary heart disease. *Medical Clinics of North America*, 1974, *58*, 269–279.

Rosenman, R. H., Brand, R. J., Jenkins, C. D., Friedman, M., Straus, R., and Wurm, M. Coronary heart disease in the Western Collaborative Group Study: Final follow-up experience of 8½ years. *Journal of the American Medical Association*, 1975, *223*, 872–877.

Scheier, M. F. Self-awareness, self-consciousness, and angry aggression. *Journal of Personality*, 1976, *44*, 627–644.

Scherwitz, L., Berton, K., and Leventhal, H. Type A behavior, self-involvement, and cardiovascular response. *Psychosomatic Medicine*, 1978a, *XL*, 593–609.

Scherwitz, L., Leventhal, H., Cleary, P., and Laman, C. Type A behavior: Consideration for risk modification. *Health Values: Achieving High-Level Wellness*, 1978b, *2*, 291–296.

Zyzanski, S. J., Jenkins, D. C., Ryan, T. J., Flessas, A., and Everist, M. Psychological correlates of coronary angiographic findings. *Archives of Internal Medicine*, 1976, *136*, 1234–1237.

Chapter 4

Sociocultural Risk Factors in Coronary Heart Disease

S. Leonard Syme And Teresa E. Seeman

During the last thirty years, research evidence has accumulated regarding the role of sociocultural factors in the etiology of arteriosclerotic cardiovascular disease. While this research evidence is now vast in quantity, it is also variable in quality. Nevertheless, an impressive set of relatively consistent findings has now emerged in this field. The purpose of this paper is to review and summarize this evidence.

Before turning to this review, it may be of interest to note some of the difficulties in conducting research on sociocultural factors. Many of our research difficulties would be solved if we could randomly allocate groups of people to differing sociocultural experiences (e.g., one group assigned to experience stressful life events while a second, "control" group is not). Instead, we must study people as we find them; people do not randomly allocate themselves to living circumstances for the sociologist's convenience. Thus, when we come upon people in various occupations, geographic areas, or behavioral categories, we can be relatively certain that they are living as they do for nonrandom reasons. One of our major challenges is to draw responsible causal inferences from observational data of nonrandom, real-life circumstances.

In addition to the problems associated with such observational data, measurement methods in this field are not as rigorous or refined as might be desired. Unlike measurement of physiological characteristics such as weight, height, and blood pressure, measurement of sociocultural characteristics frequently requires more active participation by the subject. For example, researchers must frequently rely on their subjects' ability and/or willingness to recall and report accurately past activities, thoughts, or feelings. The accuracy of self-reports varies depending on the ease with which the requested infor-

S. Leonard Syme and Teresa E. Seeman • Department of Biomedical and Environmental Health Sciences, School of Public Health, University of California at Berkeley, Berkeley, California 94720.

mation can be recalled and reported (Cannell, Marquis, and Laurent, 1977; Cannell, Miller, and Oksenberg, 1981; Balamuth, 1965; Kirscht, 1971). Information that is difficult to recall because it is from a distant past or was not a salient event for the respondent will be less accurately reported as will be relatively undesirable or embarrassing information. Additional measurement problems arise from the relative lack of consensus regarding how best to measure various sociocultural "constructs"—specifically, to identify their important dimensions and the ways in which one can obtain the most accurate information about them. As a result, a variety of tools has been used to measure particular sociocultural characteristics, and one is never sure whether contradictory results are due to "real" differences or simply to variations in measurement approaches.

In view of these problems, the likelihood is not great that a single study will produce a definitive test of an hypothesis. A more realistic expectation is that a number of studies must consistently show the same pattern of findings in regard to an hypothesis. Greater confidence in this pattern of results will exist if these research projects have been conducted in different populations using different methods. With regard to measurement methods in particular, the diversity of questionnaire items and scales in use suggests that each method may measure things somewhat differently. When we find similar results despite such diversity, confidence is increased that these measures are identifying a meaningful factor. If the same results are observed even though different researchers have used different methods among different groups of people, we might be more willing to conclude that the hypothesis is worthy of serious consideration. This view is quite different from the usual replication standards required for research. Given the special difficulties and uncertainties of research in this field, however, we may learn more through an emphasis on variation than from one on standardization.

There are three themes discernible in the research literature linking sociocultural factors to coronary heart disease: mobility, "Type A" behavior, and stressful life events. No one of the studies cited is definitive in the usual sense and, if considered alone, no one of the studies would be worthy of special attention. That a number of these studies seems to show the same pattern of findings, however, is more impressive. The three themes here are supported by relatively strong and consistent empirical evidence so that such attention seems merited.

The first of these themes concerns the relationship of various types of social and cultural mobility to coronary heart disease. In this research, it has been shown that the risk of coronary heart disease increases with major changes in place of residence. In a study in North Dakota (Syme, Hyman, and Enterline, 1964), it was found that men with two or more cross-county moves experienced approximately twice the rate of CHD observed among those with at most one such move. This same study also found that men with more major occupational changes were at greater risk: Those men with four or more major changes had about 2.5 times the rate of CHD as those who experienced 0–1 such change. A subsequent study in California (Syme, Borhani, and Buechley, 1965) confirmed this association, showing again that men with three or more occupational changes experienced twice the rate of CHD as those who were less mobile. Those who worked in three or more jobs and spent less than 30 years in their principal occupation had about four times as much CHD as those working 30+ years in their principal occupation and reporting few (0–2) job changes. Kaplan, Cassel, Tyroler, Coroni, Kleinbaum, and Hames (1971) also report that intergenerational, upward occupational mobility among the lower social classes is associated with increased rates of CHD. However, these investigators did not find such an increase for intragenerational or intergenerational mobility among the upper social classes. In the Framingham study (Haynes, Feinleib, Levine, Scotch, and Kannel, 1978; Haynes, Feinleib, and Kannel, 1980), occupational mobility as measured by number of times promoted in the past ten years was also significantly related to an increased risk of CHD among men during an eight-year follow-up, but was unrelated to initial CHD prevalence.

Other studies, though less directly concerned with occupational or geographic mobility per se, have obtained results indicating that general sociocultural mobility may be associated with greater risk of CHD. In most of these studies, discrepancies between people's current sociocultural situation and that of their childhood were found to be associated with increased rates of CHD (Jenkins, 1976; Shekelle, Ostfeld, and Paul, 1969; Christensen and Hinkle, 1961; Medalie, Kahn, Neufeld, Riss, Goldbourt, Perlstein, and Oron, 1973; Marmot and Syme, 1976; Cohen, Syme, Jenkins, Kagan, and Zyzanski, 1979). For example, the prospective study by Shekelle et al. (1969) of male employees of the Western Electric Company in Chicago found greater risk of CHD for those men whose current social class was higher or lower than that of their fathers. This study also indicated an increased risk among men with greater status inconsistency (i.e., a difference of more than three points between their highest and lowest rankings on different social status variables such as education, occupation, income, home neighborhood, and religion). The mobility studies of Syme et al. (1964, 1965) also found greater risk of CHD among men presently in white-collar jobs but who were raised on a farm or whose fathers were foreign-born. Similarly, Christensen and Hinkle (1961) reported an increased number of myocardial infarctions among former skilled craftsmen without a college education who had advanced to managerial positions as compared either with other skilled craftsmen who had not advanced or with college-educated managers.

A study by Medalie et al. (1973) also suggests the possible role of social and cultural change in increased rates of CHD. In this study, first generation Israeli civil servants had higher CHD rates than did second generation or immigrant employees. One could hypothesize that the immigrant generation retained more of the traditional culture of their country of origin, and that this orientation was in conflict with an Israeli cultural orientation. It is possible that it is within the first generation that the acculturation process of shifting to a more Israeli orientation takes place most strongly, thus placing them at greater risk of CHD. Further evidence of such an acculturation effect has also been reported from a Japanese–American study. Marmot and Syme (1976) have shown that among Japanese–American men in California, those adopting a more Western style of life (e.g., more nonJapanese associates, less frequent use of the Japanese language) had 2.5 to 5 times more CHD than did the more traditionally Japanese men. Also, among the Japanese Americans living in Hawaii, men who were characterized by both this Western acculturation and who exhibited more Western "hard-driving" behavior have been found to experience three times as much CHD as men with neither of these characteristics (Cohen *et al.*, 1979).

Other research evidence, however, has not shown an association between mobility and increased risk of CHD. As mentioned, the Framingham study showed mixed results. While number of promotions was associated with risk of CHD, other measures of occupational and social mobility (e.g., number of job changes, intergenerational, educational, and occupational mobility) were not associated with increased risk of CHD (Haynes *et al.*, 1980), and none of the mobility measures was associated with initial CHD prevalence (Haynes *et al.*, 1978). Similar negative findings have been reported from the Western Collaborative Group Study (Williams, 1968), a prospective investigation of CHD among the male employees of a group of ten California companies. For these men, no association was found between geographic or occupational mobility and CHD incidence. In a study of Bell System employees, Hinkle, Whitney, Lehman, Dunn, Benjamin, King, Plakum, and Flehinger (1968) found no evidence that occupational and/or geographic mobility (e.g., promotions and/or transfers) were associated with an increased risk of CHD. It should be noted, however, that this study population only included men who had been employed by the Bell System for five or more years, thus excluding from study the more mobile employees who might have otherwise contributed to the rate of CHD in the "mobile" group.

One explanation regarding these negative findings has been that measures of mobility in these studies reflect primarily minor degrees of occupational or social mobility, and that only major geographic, occupational, or social changes are associated with increased risk of CHD. Another possible explanation that may clarify some of these negative findings is that mobility may only function as a risk factor for CHD when it represents movement into situations with which people are unfamiliar and for which they are unprepared. As proposed by Cassel (1976), the increased rates of disease seen among the mobile and/ or those exposed to change may stem from a lack of adequate feedback in such situations, where "individuals are unfamiliar with the cues and expectations of the society in which they live." People in these situations are less able to communicate and function appropriately because they lack the knowledge and/or skills necessary for that situation; they find themselves in a situation for which they are unprepared. One result of this may be physiologic dysfunction, perhaps in the guise of CHD.

The question remains as to why and how mobility is associated with CHD. Unanswered in the mobility research is whether risk increases because of the *mobility*, because of the *situation* to which the person moves, or because of the characteristics *predisposing* certain persons to become mobile. Paralleling the stress theories of Selye, for example, it may be that any change in and of itself entails demands for adaptation which may have deleterious physiological consequences. On the other hand, mobility may be associated with CHD because of the situation into which a person moves. Thus, an unfamiliar and/ or hostile situation may increase the risks of CHD whereas movement into a "known" and/or friendly environment may not have this negative effect. An example of this might be a blue-collar *vs.* a white-collar employee moving into a new white-collar job or a member of a minority *vs.* a white, Anglo-Protestant moving into a predominantly white, Anglo-Protestant neighborhood.

Some of the negative effects of mobility may stem from enforced mobility where the individual has not chosen to make occupational, geographic, or cultural changes but rather has been constrained to make such changes (Fischer, Jackson, Stueve, Gerson, Jones with Baldssare, 1977). At least one study has reported an increase in coronary heart disease among persons living in communities undergoing social change; in this instance, increased rates of disease were observed among persons in North Carolina who had not chosen to be mobile, but who experienced the changes associated with urbanization of their environment (Tyroler and Cassel, 1964). A similar study of the health effects of social change has been reported by Suzman, Voorhees-Rosen, and Rosen (unpublished manuscript, 1980). These investigators studied the health effects of the developing North Sea oil industry in the Shetland Islands. In contrast to the North Carolina study, no negative health consequences were observed during a follow-up period of several years. In considering these contradictory results, it is noteworthy that a major difference between the populations in these two studies was the degree to which the people themselves were able to control the social change in their environment. In contrast to the North Carolina communities, the Shetlanders had a large degree of control and choice over how the oil industry would develop in their community. Most studies of mobility have not measured this "choice" aspect of change, and, therefore, there is no other evidence currently available with which to assess its possible importance. It may be that this element has important ramifications for how people experience and deal with social, geographic, and/or occupational change.

In addition to (a) mobility, (b) change itself, or (c) the situation into which a person moves, mobility may be related to CHD because of certain personal characteristics that predispose individuals both to become mobile and to develop CHD. Thus, for example, it may be that "Type A" individuals, with their greater ambition and drive, are at higher risk for both CHD and mobility. As seen in the Japanese American study, however, those Japanese who had moved to California but who retained their traditional Japanese life-style did not experience an increased rate of CHD (Marmot and Syme, 1976). Thus, it may be that such mobility only becomes a risk factor for CHD when coupled with certain other factors such as particular environmental characteristics or characteristic ways individuals have of dealing with their environment (e.g., adoption of new life-style patterns or "Type A" behavior).

Any one or all of these dimensions of mobility may contribute in different ways to coronary heart disease. Indeed, there may be multiple pathways by which mobility can influence disease—either because certain people become mobile, because certain situations are frequently associated with mobility (e.g., unfamiliarity with the setting and/or its people and their ways), and/or because of the mobility or change itself.

While the findings relative to mobility and CHD are somewhat crude and show only indirect associations, they are impressive in that investigators using different research methods in different population groups have independently come up with similar findings. In all cases where examined, the findings of increased risk were also independent of the effects of such recognized risk factors as age, sex, serum cholesterol, cigarette smoking, and blood pressure.

A second major theme that has emerged in CHD research is the relationship between the Type A behavior pattern and increased risk of coronary heart disease. Type A individuals have been characterized as competitive, aggressive, impatient, restless, and hyperalert, with tense facial musculature and "feelings of being under pressure of time and challenge of responsibility" (Zyzanski and Jenkins, 1970). During the last twenty years, Rosenman, Friedman, and colleagues have published a series of papers showing an increased rate of coronary heart disease among men characterized by such a behavior pattern (Jenkins, 1971; Jenkins, 1976; Zyzanski, 1977; Friedman and Rosenman, 1959; Friedman, Rosenman, and Carroll, 1958; Rosenman and Friedman, 1961; Rosenman, Friedman, Straus, Wurm, Jenkins, and Messinger, 1966; Rosenman, Friedman, Straus, Wurm, Kositchek, Hahn, and Werthessen, 1964; Jenkins, Zyzanski, Rosenman, and Cleveland, 1971; Rosenman, Brand, Jenkins, Friedman, Straus, and Wurm, 1975; Jenkins, Rosenman, and Friedman, 1966; Feinleib, Brand, Remington, and Zyzanski, 1977; Brand, 1977; Rosenman, Friedman, Straus, Jenkins, Zyzanski, and Wurm, 1970; Jenkins, Rosenman, and Zyzanski, 1974; Jenkins, Zyzanski, and Rosenman, 1976; Rosenman, Brand, Sholtz, and Friedman, 1976). In some of their early studies, Rosenman and Friedman compared groups of men and women characterized by Type A or Type B behavior patterns with respect to such CHD risk factors as serum cholesterol, blood clotting time, and smoking history (Friedman and Rosenman, 1959; Rosenman and Friedman, 1961). In these studies, those with Type A behavior were consistently found to have worse risk profiles: shorter clotting times, higher serum cholesterol, and more individual signs of CHD such as arcus senilus. Rosenman and Friedman also developed a structured interview to assess "Type A" behavior and successfully used this interview in their Western Collaborative Group Study. In this study, those with the Type A behavior pattern were found to have both a higher prevalence and incidence of CHD (Rosenman *et al.*, 1966; Rosenman *et al.*, 1964; Rosenman *et al.*, 1975).

Study of this behavior pattern and its relationship to CHD has also been undertaken by a large number of other investigators (Jenkins, 1971; Jenkins, 1976). In the United States, these various studies have generally confirmed the association between behavior pattern and CHD (Bruhn, Paredes, Adsett, and Wolf, 1974; Shekelle, Schoenberger, and Stamler, 1976; Glass, 1977; Kenigsberg, Zyzanski, Jenkins, Wardwell, and Licciardello, 1974; Wardwell and Bahnson, 1973). This consistency is all the more noteworthy in view of the diversity of case/control and community samples studied as well as the different methods used to assess Type A behavior such as the Structured Interview developed by Rosenman and Friedman or the Jenkins Activity Survey, a paper and pencil, self-administered questionnaire. In addition to this work, consistent findings have also been reported by investigators from the Netherlands, Belgium, Poland, and Sweden (Jenkins, 1976; Zyzanski, Wrzesniewdki, and Jenkins, 1979; Liljefors and Theorell, 1970; Theorell and Rahe, 1972; Van Dijl, 1974; Bengtsson, Hallstrom, and Tibblin, 1973; Bonami and Rime, 1972). While some of these studies did not measure the Type A behavior pattern per se, each provides evidence that various components of this behavior pattern are associated with CHD in a variety of populations, using diverse indices. Generally, these studies tend to implicate striving, diligence, ambition, job involvement, and perfectionism as risk factors for CHD. A major weakness of these studies, however, is that almost all of them are based on the study of persons who already had coronary heart disease. Given this situation, it is impossible to determine whether this behavior pattern actually predates the onset of disease or is merely a behavioral manifestation of CHD.

Important results, however, are now available from at least two prospective investigations in which the behavior pattern was assessed prior to the onset of disease. As mentioned earlier, the Western Collaborative Group Study

(WCGS) was a prospective investigation of CHD in a sample of male employees in California. After 8½ years of follow-up, the WCGS has reported a twofold excess of coronary heart disease among men with the Type A behavior pattern after account had been taken of other risk factors such as age, blood pressure, cigarette smoking, relative weight, and serum cholesterol level (Rosenman *et al.*, 1975, 1976). The second prospective study of CHD is the Framingham study, in which a community-based sample was studied. Results from this study are consistent with those from the WCGS though a different 10-item scale was used to measure the behavior pattern. This study also found that the Type A–CHD relationship seems to hold for women as well as men (Haynes *et al.*, 1980).

Although the majority of studies has found an association between Type A and CHD, some negative findings do appear. Keith, Lown, and Stare (1965) found no significant difference between their cases and the controls in prevalence of the Type A behavior pattern. Cassel (1966) and Friedman and Rosenman (1966) have each pointed out, however, that (a) the prevalence of Type A was in fact greater among the cases, and (b) that hospitalized cases may not have exhibited their true "Type A-ness" due to the hospital environment. Ahnve, DeFaire, Orth-Gomer, and Theorell (1979), using an adjective checklist to measure the behavior pattern, found no differences between three groups comprised respectively of post-infarction cases, men who had come to coronary care with chest pain but were found free of MI (myocardial infarction), and healthy male controls from three companies. Other researchers have found only partial substantiation of the Type A–CHD association. In the Japanese–American study, for example, Cohen *et al.* (1979) found an association between CHD and hard-driving, competitive Type A behavior, but only for the culturally mobile men. The Framingham study found a reversal of the usual relationship among retired blue-collar men such that the Type B men were apparently at greater risk of CHD (Haynes *et al.*, 1980) than the Type A men. Such negative or partial findings, however, appear to represent only a very small fraction of the research on Type A behavior and CHD. For the most part, Type A behavior, however measured, has been found to be a risk factor for coronary heart disease.

With the exception of the Framingham study, almost all the work on Type A and CHD has been done among employed, middle-class white men living in Western countries. It has been suggested that enhanced understanding of the behavior pattern and its relationship to CHD would be gained if they were studied among men and women living in other sociocultural circumstances and environments (Dembroski, Weiss, Shields, Haynes, and Feinleib, 1978). The analysis by Cohen *et al.* (1979) of data from the Japanese–American study indicates, in fact, that the Type A pattern as measured by the Jenkins Activity Survey (JAS) reflects a set of core items (those measuring the Hard-Driving,

Competitive components of the pattern) which are consistently related to CHD. However, unlike the usual JAS results, the Job Involvement and Hard-Working components of the scale are not related to increased risk of CHD among these Japanese–American men. These results suggest that our current assessments of Type A behavior may include extra dimensions of the pattern (e.g., Hard-Working, Job-Involved) which are actually unrelated to risk of CHD, but which show up correlated with increased CHD risk in Western populations simply because of their association in these cultures with the *core* Hard-Driving and Competitive components of the Type A pattern. In other cultures such as the Japanese, these "extra" components of the pattern may be less closely associated with the Hard-Driving and Competitive components and thus not found to be associated with CHD.

Currently, research continues regarding the component parts of this behavior pattern and how they may be related to the pathophysiology of CHD. Recent research by Glass (1977) has indicated that Type A persons may be characterized by a strong need to feel in control, suggesting that much of their striving, competition, and aggression may be directed toward achieving such control. In experimental research with Type As and Type Bs, Glass has found that when confronted by uncontrollable situations, Type As are more prone to develop a "giving-up" syndrome characterized by less effort and longer learning times on various experimental tasks (Glass, 1977). The importance of this research is twofold: (a) in our world there is much we cannot control; Type As may be faced with such situations fairly frequently, leading to the pathophysiological responses that enhance development of CHD; and (b) this research may tie in with other recent findings referred to above regarding the possible association of uncontrollable life events and disease (Streiner, Norman, McFarlan, and Roy, 1981; Fairbank and Hough, 1979).

Another component of the Type A pattern that has come under study relates to the hostility and aggression characteristic of such individuals. Various experimental studies have found that in challenge or competitive situations, Type A persons more frequently respond with hostility and/or aggressive behavior (Glass, 1977; VanEgeren, 1979). The possible importance of such a Type A characteristic in the genesis of CHD is further suggested by research on angiography patients where hostility, as measured by the MMPI, has been found to be associated with greater degree of atherosclerosis (Williams, Haney, Gentry, and Kong, 1978).

Despite our need for a better understanding of the Type A behavior pattern, it is impressive that a relatively straightforward behavioral classification of people permits the subsequent prediction of coronary heart disease independently of other recognized risk factors. The findings from two prospective studies are consistent with a much larger body of retrospective data on the association between Type A behavior and CHD; the issue now is to explore further the links between this behavior pattern or parts thereof and coronary heart disease.

A third theme now emerging with regard to sociocultural factors in CHD concerns the relationship of stressful life events to coronary heart disease. A vast literature has recently been developed showing an association between a wide variety of life events and a broad array of disease outcomes (Dohrenwend and Dohrenwend, 1974; Rahe, 1972; Goldberg and Comstock, 1976). With regard to CHD, several studies have found that myocardial infarction cases report more life events for the period just prior to their MI than do controls (Theorell and Rahe, 1971; Rahe and Paasikivi, 1971). However, virtually all of this research has been retrospective, and it is possible that those who already have disease may be more motivated than healthy persons to recall particular life events to account for their ill health (e.g., their MI). Because of such possible recall bias, the retrospective evidence seems, at best, suggestive.

Several prospective studies have been reported, but these studies deal with the incidence in young people of such relatively minor conditions as upper respiratory infections, gastrointestinal upset, and skin problems (Rahe, 1972). It may be that those who are willing to visit physicians when they have minor health complaints may also be more willing to report life event problems. One prospective study of CHD, however, has been reported by Theorell, Lind, and Floderus (1975). Measures of life events were obtained from members of a Swedish construction workers' union, and the group was subsequently monitored for illness and mortality experience. During the first twelve months of follow-up, those who developed MIs were found to have reported no more life events than had those who remained free of disease. One life event, however, that of increased work responsibility, was found to predict about 10% of the near-future MIs.

Several other prospective studies are also available now regarding the effects of particular life events such as bereavement and subsequent morbidity and mortality on CHD (Parkes, Benjamin and Fitzgerald, 1969; Jacobs and Ostfeld, 1977; Maddison and Viola, 1968; Rees and Lutkins, 1967; Lynch, 1979). Parkes *et al.* (1969), for example, found an increased rate of CHD mortality among widowers in the year following the death of their spouse. These findings are corroborated by studies such as that by Kraus and Lilienfeld (1959) which show that the divorced and widowed consistently have higher mortality rates than do the married. In addition to such "life events" research, the research findings regarding mobility (a "life event") and its possible health effects would seem to provide further evidence in favor of a role for such life events in illness onset.

While the life events approach has shown some promise in a variety of retrospective and prospective studies, it still needs to be tested rigorously in well-designed prospective studies. At the same time, research in this area will likely benefit from current efforts to improve the measurement of life events. Investigators are currently studying the different effects of positive *vs.* negative events, controllable *vs.* uncontrollable events, and anticipated *vs.* unanticipated events (Streiner *et al.,* 1981; Fairbank and Hough, 1979; Goldberg and Comstock, 1976). Such study should improve our understanding of the link between life events and illness such as CHD.

As presented above, much of the research on CHD has tended to focus on certain major risk factors of interest such as mobility, "Type A" behavior pattern, and life events. More recently, however, there has been growing interest in studying the interrelationships among such risk factors. One factor which has been proposed as a possible unifying concept for such CHD risk factors is that of interrupted social ties. It is possible that this concept may provide a parsimonious explanation for the findings summarized above. Thus, persons who have experienced occupational and residential mobility also will more likely have experienced more interrupted social relationships with other people. Certainly, this is true of those widowed. Indeed, the most important items on the stressful life events list (divorce, death of a relative) involve major interruptions of social relationships. It also may be that persons exhibiting "Type A" behavior have little time or interest in investing energy to maintain "unproductive" close ties with other people. A Belgian study of CHD (Bonami and Rime, 1972), for example, found that those men who later developed CHD reported more interest in job-related activities than in family or other social activities.

The importance of interrupted social ties was first suggested by John Cassel (1976). He argued that the common aspect of various social circumstances associated with disease (e.g., mobility and social change) was a lack of meaningful social contacts, that they reflected circumstances where individuals were less likely to receive adequate feedback as to whether their actions were producing the expected consequences. There is now a growing body of literature supporting a relationship between favorable health outcomes and the presence of meaningful social contacts in the form of social support from others (Cobb, 1976; Kaplan, Cassel, and Gore, 1977). One recent work is a longitudinal analysis of nine-year mortality in a cohort of some 6,000 adults (Berkman and Syme, 1979). In this study, an increased mortality rate was found among persons previously identified as having fewer friends and contacts with other persons. This relationship was linear and was independent of health status at baseline, as well as of other risk factors such as obesity, cigarette smoking, physical inactivity, and other health habits. Similar relationships between social support and better health have been reported by others including Nuckolls, Cassel, and Kaplan (1972); Cobb and Kasl (1977); Gore (1978); Lin, Simeone, Ensel, and Kuo (1979); LaRocco, House, and French (1980); House and Wells (1978); and House, Wells, Landerman, McMichael, and Kaplan (1979). Gore, for example, found that men experiencing unemployment but who felt support showed less increase in serum cholesterol, less depression, and reported fewer physical symptoms. The research of Nuckolls *et al.* (1972) also suggests that social support from a social network may act as a buffer, minimizing the negative consequences of various life events. In their study of pregnant women, Nuckolls *et al.* found that subsequent pregnancy complications were greatest among women who reported more life events but little perceived support (91% had complications), while women reporting similarly high numbers of life events but who also had high levels of perceived support did not show the same high prevalence of pregnancy complications (only 33% had complications). A third study by Lowenthal and Haven (1968) demonstrated that individuals with a confidante experienced less depression in dealing with widowhood than did individuals facing this same situation without a confidante.

In the area of occupational health, studies by LaRocco *et al.* (1980), House and Wells (1978), and House *et al.* (1979) have shown that social support from family, friends, and coworkers can help buffer the negative psychological and physical effects of job stress and strain. The study by House *et al.* found that such support seemed to buffer the effects of job strain on CHD risk as measured by elevated levels of two of the three major CHD risk factors—smoking, high blood pressure, and/or high serum cholesterol. Results were also presented earlier from the Japanese–American study showing that those retaining a more traditional Japanese life-style had less CHD. While the explanation for this finding is not clear, it is possible that traditional Japanese maintain more long-term, close and intimate ties with family and friends (Matsumoto, 1970).

Evidence such as this suggests that social ties may have important consequences for health and illness. It is possible that the concept of social ties may provide a unifying framework for other psychosocial risk factors such as mobility, life events, or "Type A" behavior. It may be that these latter risk factors are all associated with an increased risk of disease because they reflect a particular way of living (i.e., striving, being mobile, seeking new experience and change) that entails physiological effects which are deleterious to the human body. The importance of social ties may be that the support and/or aid they can provide may serve to buffer the negative effects of the adaptive demands placed on individuals in various situations.

A basic concept in epidemiology is that disease results from an interaction of the individual with the environment. Social ties may represent an important element influencing this person–environment interaction. They represent the links between the individual and others, influencing not only one's social interactions but also one's experience of the general environment. As exchange routes for instrumental aid as well as information, social ties are sources of both information about the meanings of events and other aspects of the environment and of information which can facilitate dealing with this environment. In this respect, social ties may represent an important link in the etiology of disease such as CHD. They may provide for healthful person–environment interaction, for the "meaningful social contacts" and "adequate feedback" John Cassel hypothesized were so important for health.

This is certainly a hypothesis worth testing in view of the fact that the recognized risk factors for coronary heart disease do not totally explain all of the disease that occurs. Data from the pooled findings of the six main prospective studies in the United States show that of men with two or more risk factors, only about 10% develop coronary heart disease over a 10-year period, while 90% do not. Viewed differently, 42% of the cases that developed in the 10-year period occurred among men without two or more of these risk factors (Intersociety Commission for Heart Disease Resources, 1970). Without de-emphasizing the undoubted importance of the recognized risk factors in the etiology of CHD, it is evident that other factors also are importantly involved. Evidence from research on social–cultural factors suggests that the concept of interrupted social ties must be considered in addition to those of mobility, life events, and "Type A" behavior.

This suggestion has the disadvantage that it proposes the addition of yet another risk factor at a time when a very large number of risk factors is already claiming our attention. Given the uncertainty of our current level of understanding, however, it seems timely to propose a new model. Such a model would take account of all important risk factors and of the way in which they are interrelated. This model would stimulate research that simultaneously examines the contribution of physiologic, psychologic, and sociocultural factors. Whether based upon laboratory investigations, cohort studies, or reanalysis of old data, such a multifactorial approach seems required to advance our understanding of arteriosclerosis.

References

Ahnve, S., DeFaire, U., Orth-Gomer, K., and Theorell, T. Type A behavior in patients with noncoronary chest pain admitted to a coronary care unit. *Journal of Psychosomatic Research*, 1979, *23*, 219–223.

Balamuth, E. Health interview responses compared with medical records. Vital and Health Statistics, Series 2, #7 1965 (National Center for Health Statistics).

Bengtsson, C., Hallstrom, T., and Tibblin, C. Social factors, stress experience and personality traits in women with ischemic heart disease, compared to a population sample of women. *Acta Scandinavica*, 1973, *549*, 82–92.

Berkman, L. F., and Syme, S. L. Social networks, host resistance and mortality: A nine-year follow-up study of Alameda Country residents. *American Journal of Epidemiology*, 1979, *109*, 186–204.

Bonami, M., and Rime, B. Approche exploratoire de la personalite precoronarienne par analyse standardisee de donnees projectives thematiques. *Journal of Psychosomatic Research,* 1972, *16*, 103–113.

Brand, R. J. Coronary-prone behavior as an independent risk factor for coronary heart disease. In T. M. Dembroski, S. M. Weiss, J. L. Shields, S. G. Haynes, and M. Feinleib (Eds.), *Coronary-Prone behavior.* New York: Springer-Verlag, 1977.

Bruhn, J. G., Paredes, A., Adsett, C. A., and Wolf, S. Psychological predictors of sudden death in myocardial infarction. *Journal of Psychosomatic Research,* 1974, *18*, 187–191.

Cannell, C., Marquis, K., and Laurent, A. A summary of studies of interviewing methodology. Vital and Health Statistics, Series 2, #69 1977 (National Center for Health Statistics).

Cannell, C., Miller, P., and Oksenberg, L. Research on interviewing techniques. In S. Leinhardt (Ed.), *Sociological methodology.* San Francisco: Jossey-Bass, 1981.

Cassel, J. C. Letter to the Editor. *Psychosomatic Medicine,* 1966, *28*, 283–284.

Cassel, J. The contribution of the social environment to host resistance. *American Journal of Epidemiology,* 1976, *104*, 107–123.

Christensen, W. N., and Hinkle, L. E. Differences in illness and prognostic signs in two groups of young men. *Journal of the American Medical Association,* 1961, *177*, 247–253.

Cobb, S. Social support as a moderator of life stress. *Psychosomatic Medicine,* 1976, *38*, 300–314.

Cobb, S., and Kasl, S. V. *Termination: The consequences of job loss.* U.S. Department of Health, Education, and Welfare. HEW (NIOSH) Publication No. 77-224. Washington, D.C.: U.S. Government Printing Office, 1977.

Cohen, J. B., Syme, S. L., Jenkins, C. D., Kagan, A., and Zyzanski, S. J. Cultural contest of Type A behavior and risk for CHD: A study of Japanese American males. *Journal of Behavioral Medicine,* 1979, *2*, 375–384.

Dembroski, T. M., Weiss, S. M., Shields, J. L., Haynes, S. G., and Feinleib, M. *Coronary-Prone Behavior.* New York: Springer-Verlag, 1978.

Dohrenwend, B. S., and Dohrenwend, B. P. *Stressful life events: Their nature and effects.* New York: Wiley-Interscience, 1974.

Durkheim, E. *Suicide—A Study in Sociology.* New York: American Book–Knickerbocker Press, 1951.

Fairbank, D. T., and Hough, R. L. Life event classifications and the event–illness relationship. *Journal of Human Stress,* 1979, *5*, 41–47.

Feinleib, M., Brand, R. J., Remington, R., and Zyzanski, S. J. Summary section: Association of coronary-prone behavior pattern and coronary heart disease. In T. M. Dembroski, S. M. Weiss, J. L. Shields, S. G. Haynes, and M. Feinleib (Eds.), *Coronary-Prone Behavior.* New York: Springer-Verlag, 1977.

Fischer, C. S., Jackson, R. M., Stueve, C. A., Gerson, K., Jones, L. M., with Baldssare, M. *Networks and places: Social relations in the urban setting.* New York: Free Press, 1977.

Friedman, M., and Rosenman, R. H. Association of specific overt behavior pattern with blood and cardiovascular findings. *Journal of the American Medical Association,* 1959, *169*, 1286–1296.

Friedman, M., and Rosenman, R. H. Letter to the Editor. *Psychosomatic Medicine,* 1966, *28*, 282.

Friedman, R., Rosenman, R. H., and Carroll, V. Changes in the serum cholesterol and blood clotting time in man subjected to cyclic variation in occupational stress. *Circulation,* 1958, *17*, 852–861.

Glass, D. C. Stress, behavior pattern, and coronary disease. *American Scientist,* 1977, *65*, 177–187.

Goldberg, E. L., and Comstock, G. W. Life events and subsequent illnesses. *American Journal of Epidemiology,* 1976, *104*, 146–158.

Gore, S. The effect of social support in moderating the health consequences of unemployment. *Journal of Health and Social Behavior,* 1978, *19*, 157–165.

Haynes, S. G., Feinleib, M., and Kannel, W. B. The relationship of psychosocial factors to coronary heart disease in the Framingham Study: III. Eight-year incidence of coronary heart disease. *American Journal of Epidemiology,* 1980, *III*, 37–58.

Haynes, S. G., Feinlaub, M., Levine, S., Scotch, N., and Kannel, W. B. The relationship of psychosocial factors to coronary heart disease in the Framingham Study: II. Prevalence of coronary heart disease. *American Journal of Epidemiology,* 1978, *107*, 374–402.

Hinkle, L. E., Jr., Whitney, L. H., Lehman, E. W., Dunn, J., Benjamin, B., King, R., Plakum, A., and Flehinger, B. Occupation, education, and coronary heart disease: Risk is influenced more by education and background than by occupational experiences, in the Bell system. *Science,* 1968, *161*, 238–246.

House, J. S., and Wells, J. A. Occupational stress, social support, and health. In A. McLean, C. Black, and M. Colligan (Eds.), *Reducing occupational stress: Proceedings of a conference.* U.S. Department of Health, Education and Welfare, HEW (NIOSH) Publication No. 78-140. Washington, D.C.: U.S. Government Printing Office, 1978.

House, J., Wells, J., Landerman, L., McMichael, A., Kaplan, B. Occupational stress and health among factory workers. *Journal of Health and Social Behavior,* 1979, *20*, 139–160.

The Inter-Society Commission for Heart Disease Resources. Primary Prevention of the atherosclerotic diseases. *Circulation,* 1972, *42*, 1–44.

Jacobs, S., and Ostfeld, A. An epidemiological review of the mortality of bereavement. *Journal of Psychosomatic Medicine,* 1977, *39*, 344–357.

Jenkins, C. D. Psychologic and social precursors of coronary disease. *New England Journal of Medicine,* 1971, *284*, 244–255; 307–317.

Jenkins, C. D. Recent evidence supporting psychologic and social risk factors for coronary heart disease. *New England Journal of Medicine,* 1976, 987–994; 1033–1038.

Jenkins, C. D., Rosenman, R. H., and Friedman, M. Components of the coronary-prone behavior pattern: Their relation to silent MI and blood lipids. *Journal of Chronic Diseases,* 1966, *19*, 599–609.

Jenkins, C. D., Rosenman, R. H., and Zyzanski, S. J. Prediction of clinical coronary heart disease by a test for the coronary-prone behavior pattern. *New England Journal of Medicine,* 1974, *290*, 1271–1275.

Jenkins, C. D., Zyzanski, S. J., and Rosenman, R. H. Risk of new myocardial infarction in the middle-aged man with manifest coronary heart disease. *Circulation,* 1976, *53*, 342–347.

Jenkins, C. D., Zyzanski, S. J., Rosenman, R. H., and Cleveland, C. L. Association of coronary-prone behavior scores with recurrence of coronary heart disease. *Journal of Chronic Diseases,* 1971, *24*, 601–611.

Kaplan, B. H., Cassel, J. C., and Gore, S. Social support and health. *Medical Care,* 1977, *15*, 47–58.

Kaplan, B. H., Cassel, J. C., Tyroler, H. A., Coroni, J. C., Kleinbaum, D. G., and Hames, C. G. Occupational mobility and coronary heart disease. *Archives of Internal Medicine,* 1971, *128*, 938–942.

Keith, R. A., Lown, B., and Stare, F. J. Coronary heart disease and behavior patterns: An examination of method. *Psychosomatic Medicine,* 1965, *27*, 424–434.

Kenigsberg, D., Zyzanski, S. J., Jenkins, C. D., Wardwell, W. I., and Licciardello, A. T. The coronary-prone behavior pattern in hospitalized patients with and without coronary heart disease. *Psychosomatic Medicine,* 1974, *36*, 351–444.

Kirscht, J. P. Social and psychological problems of surveys on health and illness. *Social Science and Medicine,* 1971, *5*, 519–526.

Kraus, A. S., and Lilienfeld, A. M. Some epidemiologic aspects of the high mortality rate in the young widowed group. *Journal of Chronic Diseases,* 1959, *10*, 207–217.

LaRocco, J. M., House, J. S., and French, J. R. P., Jr. Social support, occupational stress, and health. *Journal of Health and Social Behavior,* 1980, *21,* 202–218.

Liljefors, J., and Theorell, T. An identical twin study of psycho-social factors in coronary heart disease in Sweden. *Psychosomatic Medicine,* 1970, *32,* 523–542.

Lin, N., Someone, R. S., Ensel, W. M., and Kuo, W. Social support, stressful life events, and illness: A model and an empirical test. *Journal of Health and Social Behavior,* 1979, *20,* 108–119.

Lowenthal, M. F., and Haven, C. Interaction and adaptation: Intimacy as a critical variable. *American Sociological Review,* 1968, *33,* 20–30.

Lynch, J. J. *The Broken Heart: The Medical Consequences of Loneliness.* New York: Basic Books, 1979.

Maddison, D., and Viola, A. The health of widows in the year following bereavement. *Journal of Psychosomatic Research,* 1968, *12,* 297–306.

Marmot, M. G., and Syme, S. L. Acculturation and coronary heart disease in Japanese-Americans. *American Journal of Epidemiology, 104,* 1976, 225–247.

Matsumoto, Y. S. Social stress and coronary heart disease in Japan: A hypothesis. *Milbank Memorial Fund Q,* 1970, *48,* 9–36.

Medalie, J. H., Kahn, H. A., Neufeld, H. N., Riss, E., Goldbourt, U., Peristein, T., and Oron, D. Myocardial infarction over a five-year period: I. Prevalence, incidence, and mortality experience. *Journal of Chronic Disorders,* 1973, *26,* 63–84.

Nuckolls, K. B., Cassel, J. C., and Kaplan, B. H. Psychosocial asset, life crisis, and the prognosis of pregnancy. *American Journal of Epidemiology,* 1972, *95,* 431–441.

Parkes, C. M., Benjamin B., and Fitzgerald, R. G. Broken heart: A statistical study of increased mortality among widowers. *British Medical Journal,* 1969, *1,* 740–743.

Rahe, R. Subjects' recent life changes and their near-future illness reports: A review. *Annals of Clinical Research,* 1972, *4,* 250–265.

Rahe, R. H., and Paasikivi, J. Psychosocial factors and myocardia infarction—II. An outpatient study in Sweden. *Journal of Psychosomatic Research,* 1971, *15,* 33–39.

Rees, W. D., and Lutkins, S. J. Mortality of bereavement. *British Medical Journal,* 1967, *4,* 13–16.

Report of the Inter-Society Commission for Heart Disease Resources: Primary prevention of the atherosclerotic diseases. *Circulation,* 1970, *42,* A55–A95.

Rosenman, R. H., Brand, R. J., Jenkins, C. D., Friedman, M., Straus, R., and Wurm, M. Coronary heart disease in the Western Collaborative Group Study: Final follow-up experience of 8.5 years. *Journal of the American Medical Association,* 1975, *233,* 872–877.

Rosenman, R. H., Brand, R. J., Sholtz, R. I., and Friedman, M. Multivariate prediction of coronary heart disease during 8.5 year follow-up in the Western Collaborative Group Study. *American Journal of Cardiology,* 1976, *37,* 903–910.

Rosenman, R. H., and Friedman, M. Association of specific behavior pattern in women with blood and cardiovascular findings. *Circulation,* 1961, *24,* 1173–1184.

Rosenman, R. H., Friedman, M., Straus, R., Wurm, M., Jenkins, C. D., and Messinger, H. B. Coronary heart disease in the Western Collaborative Group Study: A follow-up experience of two years. *Journal of the American Medical Society,* 1966, *195,* 130–136.

Rosenman, R. H., Friedman, M., Straus, R., Wurm, M., Kositchek, R., Hahn, W., and Werthessen, N. A predictive study of coronary heart disease. *Journal of the American Medical Society,* 1964, *189,* 15–22.

Rosenman, R. H., Friedman, M., Strauss, R., Jenkins, C. D., Zyzanski, S. J., and Wurm, M. Coronary heart disease in the Western Collaborative Group Study: A follow-up experience of 4.5 years. *Journal of Chronic Diseases,* 1970, *23,* 173–190.

Shekelle, R. B., Ostfeld, A. M., and Paul, O. Social status and incidence of coronary heart disease. *Journal of Chronic Diseases,* 1969, *22,* 381–394.

Shekelle, R. B., Schoenberger, J. A., and Stamler, J. Correlates of the JAS Type A behavior pattern score. *Journal of Chronic Diseases,* 1976, *29,* 381–394.

Streiner, D. L., Norman, G. R., McFarlan, A. H., and Roy, R. G. Quality of life events and their relationship to strain. *Schizophrenia Bulletin,* 1981, *7,* 34–41.

Suzman, R. M., Voorhees-Rosen, D. J., and Rosen, D. H. The impact of the North Sea oil development on mental and physical health: A longitudinal study of the consequences of economic boom and rapid social change. Unpublished manuscript, Langley Porter Neuro-Psychiatric Institute, University of California, San Francisco, 1980.

Syme, S. L., Borhani, N. O., and Buechley, R. W. Cultural mobility and coronary heart disease in an urban area. *American Journal of Epidemiology,* 1965, *82,* 334–346.

Syme, S. L., Hyman, M. M., and Enterline, P. E. Some social and cultural factors associated with the occurrence of coronary heart disease. *Journal of Chronic Diseases,* 1964, *17,* 277–289.

Theorell, T., Lind, E., and Floderus, B. The relationship of disturbing life-changes and emotions to the early development of myocardial infarction and other serious illnesses. *International Journal of Epidemiology,* 1975, *4,* 281–293.

Theorell, T., and Rahe, R. H. Psychosocial factors and myocardial infarction—I. An inpatient study in Sweden. *Journal of Psychosomatic Research,* 1971, *15,* 25–31.

Theorell, T., and Rahe, R. H. Behavior and life satisfactions: Characteristics of Swedish subjects with myocardial infarction. *Journal of Chronic Diseases,* 1972, *25,* 139–147.

Tyroler, J. A., and Cassel, J. Health consequences of culture change: II. Effect of urbanization of coronary heart mortality in rural residents. *Journal of Chronic Diseases,* 1964, *17,* 167–177.

Van Dijl, H. Activity and job-responsibility as measured by judgment behavior in myocardial infarction patients. *Psychotherapy and Psychosomatics,* 1974, *24,* 126–128.

Van Egeren, L. F. Social interactions, communications, and the coronary-prone behavior pattern: A psychological study. *Psychosomatic Medicine,* 1979, *41,* 2–18.

Wardwell, W. I., and Bahnson, C. B. Behavioral variables and myocardial infarction in the Southeastern Connecticut heart study. *Journal of Chronic Diseases,* 1973, *26,* 447–461.

Williams, C. A. The relationship of occupational change to blood pressure, serum cholesterol, and specific overt behavior patterns. Unpublished Ph.D. dissertation. University of North Carolina, Chapel Hill, 1968. Cited in Jenkins (Ref. 10).

Williams, R. B., Haney, T., Gentry, W. D., and Kong, Y. Relation between hostility and arteriographically documented coronary atherosclerosis. Paper presented at the American Psychosomatic Society meetings, Washington, D.C., 1978.

Zyzanski, S. J. Coronary-prone behavior pattern and coronary heart disease: Epidemiological evidence. In T. M. Dembroski, S. M. Weiss, J. L. Shields, S. G. Haynes, and M. Feinleib (Eds.), *Coronary-Prone Behavior.* New York: Springer-Verlag, 1977.

Zyzanski, S. J., and Jenkins, C. D. Basic dimensions within the coronary-prone behavior pattern. *Journal of Chronic Diseases,* 1970, *22,* 781–795.

Zyzanski, S. J., Wrzesniewdki, K., and Jenkins, C. D. Cross-cultural validation of the coronary-prone behavior pattern. *Social Science and Medicine,* 1979, *13a,* 405–412.

Chapter 5

Coronary Heart Disease and Sudden Death

Robert S. Eliot

At this time in history it is important to recognize that the capricious link between coronary atherosclerosis and clinical events has created a considerable degree of controversy for valid clinical researchers as well as basic science investigators concerning the underlying pathophysiology of myocardial infarction and sudden death. With regard to myocardial infarction, it is perplexing to discover that in one study 37% of coroners' cases of individuals who had never experienced coronary heart disease and died of other unrelated causes manifested one or major coronary arterial occlusions at autopsy (Baroldi, Falzi, and Mariani, 1979). Thus, a pool of individuals is to be found within society who live out normal life expectancies devoid of subjective or objective clinical evidence of ischemic heart disease yet possess one or more major coronary occlusions.

Another major issue is the failure to recognize the multiplicity of factors that singly or in concert influence sudden, unexpected "coronary" death. In this situation one may find sudden, disorganized ventricular rhythm disturbances, vagotonic deaths of the diving reflex variety as described by Engel (1978), evidence of early myocardial infarction, or the actual rupture of myocardial fibrils (coagulative myocytolysis).

It would be foolish to indict the neuroendocrine mechanism as the solitary cause for any or all of the aforementioned events. Nevertheless, there is a body of evidence from laboratory and clinical investigations which consistently points to interactions between the neuroendocrine system and other biochemical and physiological systems. These systems appear to interact as synergistic accomplices in producing potentially or actually fatal outcomes.

Robert S. Eliot • Department of Preventive and Stress Medicine, University of Nebraska Medical Center, Omaha, Nebraska 68105.

1. Myocardial Infarction

There is little question that environmental and behavioral factors interact in the mosaic of events associated with hypertension as described by Irving Page (1949). Interaction of these factors in hypertension is substantiated in both animal and human studies. The acute elevation of both systolic and diastolic blood pressure under circumstances of emotional stress when superimposed on an already diseased coronary arterial system may, in fact, produce or augment adverse myocardial oxygen demands by influencing one or more of the three major determinants of myocardial oxygen consumption: heart rate, wall tension, and contractile state.

Indeed, myocardial infarction and ischemic heart disease are known to occur in the absence of any significant coronary atherosclerosis. Original clinical observations were coincidentally reported by Likoff (Likoff, Segal, and Kasparian, 1967; Likoff, 1971) and Eliot (Eliot and Bratt, 1969; Eliot, Baroldi, and Leone, 1974). The postmortem evidence that this occurs is found in a study reported by Eliot *et al.* (1974). Currently, the suspected mechanisms under discussion have emphasized spasm. Evidence that ergonovine, methylcholine, or epinephrine plus propranolol can induce spasm in susceptible patients suggests that the neuroendocrine system in this instance may be involved. Chahine, Zacca, and Verani (1979) and Oliva and Breckenridge (1977) have also demonstrated the presence of spasm during acute myocardial infarction. It has been suggested that spasm may even play a role in the precipitation of the acute coronary events, and that what we are seeing is only one end of the bell-shaped curve of pathophysiologic interaction between spasm, ischemia, and infarction.

The importance of coronary thrombosis as an initiating mechanism in myocardial infarction has been widely discussed and remains controversial. Whether thrombosis is a primary or secondary event is of less importance for the purpose of this discussion than understanding the potential role which the neuroendocrine system may play in increasing platelet adhesiveness. Such investigators as Ross and Harker (1976) and others agree that this mechanism is a potentially important factor in atherosclerotic sequelae. Evidence for increased platelet adhesiveness has been observed by a host of investigators under a variety of circumstances in both animals and man in stressful situations. Unquestionably the substrate plays a key role, in that the most frequent site of platelet aggregation is the area of greatest atherosclerotic involvement and the point of maximum narrowing. Here, the interaction of lipids and/or catecholamines, which is said to increase platelet adhesiveness, and the recently discovered platelet derived growth factor, which potentially has the capability of stimulating intimal site proliferation (Kaplan, Chao, Stiles, Antoniades, and

Scher, 1979), may be synergistic mechanisms which accomplish or enhance the risk of occlusion at a point of maximum turbulent flow.

In the circumstance of coronary heart disease without infarction, namely angina, evidence from psychological studies suggests that these patients, even prior to incurring angina pectoris, are prone to develop higher states of anxiety (Buell and Eliot, 1979). It is also argued that Type A behavior, a component of coronary-prone behavior, may be a contributing factor to the emergence of coronary heart disease and, perhaps, even has something to do with development of coronary atherosclerosis. These evidences were accepted at a recent conference conducted by the National Heart, Lung, and Blood Institute and are soon to be published (The Review Panel on Coronary Prone Behavior). It is well agreed, however, that the major evidence that Type A behavior is a factor equal in stature to other identifiable coronary risk factors rests on epidemiologic studies. At the moment, it is not prognostically specific at the individual level in that it lacks adequate selectivity and specificity.

The most consistent location for early myocardial necrosis is in the inner third (subendocardium) of the left ventricular wall with or without coronary occlusion. The earliest features of the coagulation necrosis process include elongation and thinning of muscle fibers particularly with distortion of the nucleus (Figure 1). These features are distinct from those seen in catecholamine-induced necrosis which will be discussed in the next section.

2. Sudden Death

Lown, Verrier, and Rabinowitz (1977) have demonstrated that the threshold for ventricular fibrillation in both animals and man can be lowered by situations conducive to central nervous system arousal, and the supposition that the neuroendocrine system is called into action is obvious. Indeed, many of the agents used to control potentially fatal rhythm disturbances are designed to blunt neuroendocrine responses and have been demonstrated to do so in both animal and clinical studies.

Additional provocative observations have been made from coroners' studies. Baroldi (1975) has reported the presence of ruptured myocardial fibers scattered throughout the myocardium in approximately 76% of individuals who died without premonitory signs of ischemia. In such cases, there is histologic evidence of coagulation beginning near the intercalated disc and extending, in some instances, throughout the entire cell. The cell demonstrates an acidophilic response on standard H & E staining and requires only light microscopy for identification (Figure 2). In our laboratory, we have been able to duplicate these observations by injecting animals with boluses of catecholamines (Eliot,

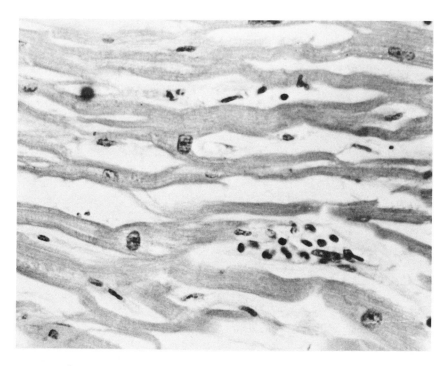

Figure 1. Coagulation necrosis. Early features of this process include elongation and thinning of muscle fibers.

Clayton, Pieper, and Todd, 1977). By employing elaborate techniques, we have also noted that the quantity of these lesions is dependent upon the catecholamine dose, and the likelihood of irreversible ventricular rhythm disturbances is dependent upon the number of lesions present. Further, as with ischemic necrosis, the inner third of the left ventricle is most vulnerable to this form of necrosis. Of particular interest is the evidence that these lesions can be sharply reduced, if not almost completely eliminated, by pretreatment with beta blockade (Eliot, Todd, Clayton, and Pieper, 1978). This type of necrosis is distinctly different from that seen with myocardial infarction in that within a period of 24 hours, if the animal can be kept alive, no coagulative myocytolysis remains. Instead, one finds the evidence of empty sarcolemmal tubes (Figure 3).

Thus, at this moment of our understanding, the term coagulative myocytolysis appears appropriate to identify the pathophysiologic mechanism, and it appears that catecholamines are operative in inducing this form of necrosis within a matter of minutes. Undoubtedly, this condition cannot be completely

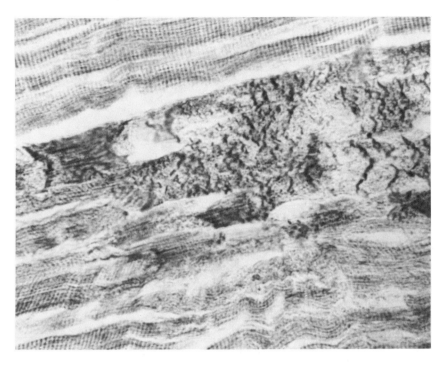

Figure 2. Coagulative myocytolysis. This lesion may be induced by catecholamine administration. It is characterized by the presence of anomalous contraction bands (hypercontraction of myofibrils).

separated from the lowered ventricular threshold for irritability as described by Lown, DeSilva, and Lenson (1978), although their reports do not yet include histopathologic data. It is also true that in sudden death following infarction one can occasionally find a large number of similar lesions, namely hyperfunctional necrosis or coagulative myocytolysis (Figure 2). The suggestion is that this form of necrosis, induced by sudden outpourings of catecholamines, may influence the ischemic myocardium. Again, the interplay between two or more different pathophysiologic states may combine and lead to potentially fatal results.

A final, well recognized mechanism of sudden death is that of ischemia itself in which rhythm and conduction disturbances can be brought about by the development of variable rates of ventricular conduction, and recovery induced by the ischemia alone. These changes are well known to enhance the possibility of reentry and reentry is a well known mechanism for ventricular tachycardia and/or fibrillation.

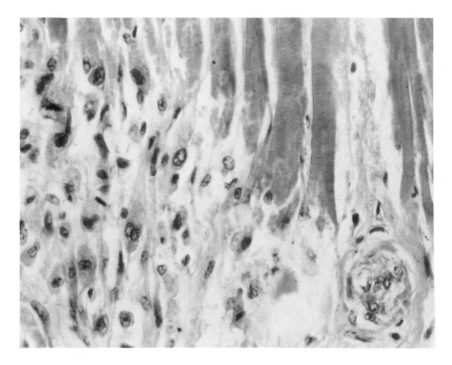

Figure 3. Empty sarcolemmal tubes are the only remaining evidence of coagulative myocytolysis after 24 hours.

3. Prospects for Future Research

The major challenge and responsibility of those who attempt to define the roles of the neuroendocrine system, behavior, and related issues in coronary heart disease is to demonstrate carefully and thoroughly that psychobiological events can be quantitated in a meaningful fashion and studied objectively. One method we are exploring is that of psychophysiological testing. Standard hemodynamic responses, including the important dimensions of cardiac output, total systemic resistance, and indices of the major determinants of myocardial oxygen requirements, are studied by noninvasive techniques during standardized psychological input.

It is vital to keep in mind that the individual perception of stress varies from one person to another as attested to in the old adage, "one man's meat is another man's poison." For this reason, it is impossible to designate circumstances, environmental conditions, or overt behavior as immutably good or bad on a general basis. This decision has to be made by the individual in question somewhere between the higher cortical centers, the limbic system, and the hypothalamus. In order to provide this opportunity and to record the results objectively, a standardized simulation of emotionally stressful tasks has been devised. The results and measures are those of the degree of physiological change, the rate of the change, and the rate of return to baseline levels as reflectors of psychophysiologic strain responses. Hopefully, through this system, along with others, we will be able to measure more objectively the influences of behavior, perception, and environment which will facilitate more selective prognostic abilities and better interventive strategies.

Given our current, limited state of understanding, large epidemiologic studies designed to measure behavior would appear to be premature. Should there be some neuroendocrine components considered worth measuring in the future, it would appear most cost-effective to include these components in other ongoing epidemiologic studies.

References

Baroldi, G. Different types of myocardial necrosis in coronary heart disease. A pathophysiologic review of their functional significance. *American Heart Journal,* 1975, *89,* 742–752.

Baroldi, G., Falzi, G., and Mariani, F. Sudden coronary death. A post mortem study in 208 selected cases compared to 97 "control" subjects. *American Heart Journal,* 1979, *98,* 20–31.

Buell, J. C., and Eliot, R. S. Stress and cardiovascular disease. *Modern Concepts of Cardiovascular Disease,* 1979, *48,* 19–24.

Chahine, R. A., Zacca, N., and Verani, M. S. Update on coronary artery spasm. *Practical Cardiology,* 1979, *5,* 27–32.

Coronary-prone behavior and coronary heart disease: A critical review. The Review Panel on Coronary-Prone Behavior and Coronary Heart Disease. *Circulation,* 1981, *63,* 1199–1215.

Eliot, R. S., Baroldi, G., and Leone, A. Necropsy studies in myocardial infarction with minimal or no coronary luminal reduction due to atherosclerosis. *Circulation,* 1974, *49,* 1127–1131.

Eliot, R. S., and Bratt, G. T. The paradox of myocardial ischemia and necrosis in young women with normal coronary arteriograms relationship to anomalous hemoglobin-oxygen dissociation. *American Journal of Cardiology,* 1969, *23,* 633.

Eliot, R. S., Clayton, F. C., Pieper, G. M., and Todd, G. L. Influence of environmental stress on the pathogenesis of sudden cardiac death. *Federation Proceedings,* 1977, *36,* 1719–1724.

Eliot, R. S., Todd, G. L., Clayton, F. C., and Pieper, G. M. Experimental catecholamine-induced acute myocardial necrosis. In V. Manninen and P. I. Halonen (Eds.), *Advances in cardiology,* Vol. 2. Basel: S. Karger AG, 1978.

Engel, G. L. Psychologic stress, vasodepressor (vasovagal) syncope and sudden death. *Annals of Internal Medicine,* 1978, *89,* 403–412.

Kaplan, D. R., Chao, F. C., Stiles, C. D., Antoniades, H. N., and Scher, C. D. Platelet diphagranule contain a growth factor for fibroblasts. *Blood,* 1979, *53,* 1043–1052.

Likoff, W. Myocardial infarction in subjects with normal coronary arteriograms. *American Journal of Cardiology,* 1971, *28,* 742–743.

Likoff, W., Segal, B. L., and Kasparian, H. Paradox of normal selective coronary arteriograms in patients considered to have unmistakable coronary heart disease. *New England Journal of Medicine,* 1967, *276,* 1063.

Lown, B., DeSilva, R. A., and Lenson R. Roles of psychologic stress and autonomic nervous system changes in provocation of ventricular premature complexes. *American Journal of Cardiology,* 1978, *41*, 979–985.

Lown, B. L., Verrier, R. L., and Rabinowitz, S. H. Neural and psychologic mechanisms and the problem of sudden cardiac death. *American Journal of Cardiology,* 1977, *39*, 890–902.

Oliva, P. B., and Breckenridge, J. C. Acute myocardial infarction with normal and near normal coronary arteries. *American Journal of Cardiology,* 1977, *40*, 1000–1007.

Page, I. H. Pathogenesis of arterial hypertension. *Journal of the American Medical Association,* 1949, *140*, 451–457.

Ross, R., and Harker, L. Hyperlipidemia and atherosclerosis. *Science,* 1976, *193*, 1094–1100.

Part III

Diagnosis, Treatment, and Rehabilitation

Overview

Biobehavioral Factors in the Clinical Management of Arteriosclerotic Cardiovascular Disease

Redford B. Williams, Jr.

Behavioral, psychological, and socio-cultural factors deserve careful attention in consideration of key issues relating to the clinical management of and rehabilitation from arteriosclerotic cardiovascular disease. Principles of behavior analysis have potentially important implications for a wide variety of treatment and rehabilitation concerns, including modification of lifestyles, compliance with medical regimens, and direct treatment of such symptoms and signs as angina and high blood pressure. Identification of psychosocial or behavioral characteristics which identify subgroups of patients with greater likelihood of survival or pain reduction would enhance rational decision making concerning surgical versus medical management treatment for a given patient. Finally, evidence is accumulating that behavioral and psychosocial characteristics of patients are in many cases as important as physiologic status in determining successful rehabilitation.

An extensive body of behavioral science research over the past four decades has helped to identify a number of basic scientifically verifiable principles governing interactions between the organism and the environment (Bandura, 1969; Brady, 1966; Brady, 1979; Miller, 1969). It is important to distinguish these principles which focus primarily on overt, observable behavior from other approaches such as psychoanalysis, which focus primarily on inferred but unobserved mental processes.

One such basic principle concerns the relationship between an observed behavior and the schedule of reinforcement serving to maintain that behavior. This principle states that the more quickly a positive consequence or reward

---

Redford B. Williams, Jr.   •   Duke University Medical Center, Department of Psychiatry, Box 3416, Durham, North Carolina 27710.

follows a given behavior, the more likely will be the repeated occurrence of that behavior. This principle has obvious implications for many important issues in health education and compliance with medical regimens. For example, if the stated reward of continuing a program of regular exercise is that 15 years later one will be less likely to suffer a myocardial infarction, it is easy to see that we are dealing with a reinforcer which is long delayed with regard to the specific behavior it is intended to reinforce. If the consequences of cigarette smoking are presented as an increased likelihood of lung cancer or myocardial infarction some 20 years hence, it is easy to see how such a reinforcer might have considerably less influence over behavior than would such immediate positive reinforcers as relief of symptoms of nicotine withdrawal, reduction of stress and an enhanced sense of togetherness with others—all of which have been reported by smokers as pleasant aspects of cigarette smoking.

Application of such principles of behavior analysis as described above has led to the conclusion (Evans, Hill, Raines, and Henderson, in press) that rather than focusing our efforts on trying to overcome the immediate positive reinforcers for those already addicted to nicotine, our efforts should focus on preventing teenagers from ever starting to smoke (see Chapter 2). Currently research in this area is proceeding at a rapid pace and it is hoped evaluation of this research will confirm that it is possible by appropriate application of behavioral science knowledge and techniques to prevent teenagers from ever taking up the cigarette habit.

With regard to treatment of patients who already have clinically evident coronary heart disease, the critical issues are to relieve angina pain and to reduce mortality. At present the key treatment choice is that between medical management (propanolol, nitroglycerin, antiarrhythmics) and saphenous vein aortocoronary bypass surgery. A number of studies (Whalen, Wallace, McNeer, Rosati, and Lee, 1977; Hultgren, Takaro, Detre, and Murphy, 1978; Kloster, Kremkau, Ritzman, Rahimtoola, Rosch, and Kanarek, 1979; Hammermeister, DeRouen, and Dodge, 1979; Harris, Phil, Harrell, Lee, Behar, and Rosati, 1979) have found that, except for patients with lesions of the left main coronary artery, prognosis for survival is no different with medical versus surgical management, especially with control for differential proportions in each treatment group with impaired left ventricular function. What is clear from these studies is that surgical treatment is superior to medical treatment with regard to relief of anginal pain. Thus, except for the approximately 10% of patients with atherosclerotic involvement of the left main coronary artery, the main indication for surgical treatment at present is to achieve relief of anginal pain.

Currently there are few reliable physiologic factors which predict with any degree of certainty who is likely to achieve significant reduction of angina with either medical or surgical treatment. Since the phenomenon of pain has long been known to have a significant psychological/behavioral component, it seems promising that the application of behavioral science knowledge and techniques could help to identify subgroups of patients more likely to obtain pain relief with either of the two currently available forms of treatment. For example, in a study currently under way at Duke, an extensive array of psychosocial and behavioral variables has been assessed on over 1300 patients referred for diagnostic coronary arteriography. Preliminary, unpublished analyses of these data suggest that there are psychosocial predictors which are strongly related to both the pain and survival outcomes independently of anatomic and hemodynamic factors. With the annual cost of coronary bypass graft surgery currently over $1,000,000,000 in the U.S., any improvement in our ability to identify prospective patients who will achieve relief of angina equally as well with medical as with surgical management could result in a dramatic saving in treatment costs.

In the area of rehabilitation, or secondary prevention, there are no reliable physiological predictors of who will subsequently comply or fail to comply with rehabilitation regimens following myocardial infarction. Recent research (Oldridge, Wicks, Hanley, Sutton, and Jones, 1978), however, has identified psychological and behavioral characteristics which appear to predict compliance with medical regimens. In addition, there are psychological and behavioral characteristics which have been found related to outcomes of the rehabilitation process: patients who were depressed prior to entry in a rehabilitation program were found on follow-up to have poor social readjustments, lower rates of return to work and greater requirements for medical care (Stern, Pascale, and Ackerman, 1977). In addition, patients who had had a previous myocardial infarction and who were Type A were significantly more likely on follow-up in the Western Collaborative Group Study to suffer a reinfarction than were Type B men (Jenkins, Zyzanski, and Rosenman, 1976).

Although it is difficult to predict which patients will have a good response to risk reduction and cardiac rehabilitation programs, many patients do make substantial improvements. Reduction of blood pressure, decrease in body weight, and physical conditioning through regular exercise all tend to reduce myocardial oxygen requirements. Also, reduction in cigarette smoking, psychological depression, and hyperlipidemia lessen the possibility of future reinfarction. All these measures require cooperation of patients in adhering to the prescribed medical regimens. Consequently, medical management of patients with arteriosclerotic cardiovascular disease requires careful attention to behavioral as well as physiological and pathological characteristics.

References

Bandura, A. *Principles of behavioral modification.* New York: Holt, Rinehart, and Winston, 1969.

Brady, J. V. Operant methodology and the production of altered physiological states. In W. Honig (Ed.), *Operant behavior: Areas of research and application.* New York: Appleton-Century-Crofts, 1966.

Brady, J. V. Learning and conditioning. In O. F. Pomerleau and J. P. Brady (Eds.), *Behavioral medicine: Theory and practice.* Baltimore: Williams and Wilkins, 1979.

Evans, R. I., Hill, P. C., Raines, B. E., and Henderson, A. H. Current psychological, social, and educational programs in control and prevention of smoking: A critical methodological review. *Atherosclerosis Reviews,* 1979, *6,* 203–245.

Hammermeister, K. E., DeRouen, T. A., and Dodge, H. T. Variables predictive of survival in patients with coronary disease. *Circulation,* 1979, *59,* 421–430.

Harris, P. J., Phil, D., Harrell, F. E., Jr., Lee, K. L., Behar, V. S., Rosati, R. A. Survival in medically-treated coronary artery disease. *Circulation,* 1979 *60,* 1259–1269.

Haynes, R. B., Taylor, D. W., and Sackett, D. L. *Compliance in health care.* Baltimore: Johns Hopkins, 1979.

Hultgren, H. N., Takaro, T., Detre, K. M., and Murphy, M. L. Aortocoronary-Artery-Bypass Assessment after 13 Years. *Journal of the American Medical Association,* 1978, *240,* 1353–1354.

Jenkins, D., Zyzanski, S. J., and Rosenman, R. H. Risk of new myocardial infarction in middle aged men with manifest coronary heart disease. *Circulation,* 1976, *53,* 342–347.

Kloster, F. E., Kremkau, E. L., Ritzman, L. W., Rahimtoola, S. H., Rosch, J., and Kanarek, P. H. Coronary bypass for stable angina. *New England Journal of Medicine,* 1979, *300,* 149–157.

Miller, N. E. Learning of visceral and glandular responses. *Science,* 1969, *163,* 434–445.

Oldridge, M. B., Wicks, J. R., Hanley, C., Sutton, J. R., and Jones, N. L. Noncompliance in an exercise rehabilitation program for men who have suffered a myocardial infarction. *Canadian Medical Association Journal,* 1978, *118,* 361–364.

Stern, M. J., Pascale, L., and Ackerman, A. *Archives of Internal Medicine,* 1977, *137,* 1680–1685.

Whalen, R. E., Wallace, A. G., McNeer, J. F., Rosati, R. A., Lee, K. L. The natural history of coronary artery disease: An update on surgical and medical management. *Transactions of the American Clinical and Climatological Association,* 1977, *89,* 19–33.

Chapter 6

Behavior Analysis and Cardiovascular Risk Factors

Joseph V. Brady

The observation that physiological functions can be altered, often in profound and enduring ways by environmental circumstances and behavioral interactions, has been repeatedly confirmed in both the laboratory and the clinic. The current resurgence of interest in the analysis of these relationships and their application in behavioral medicine reflects the strong empirical and conceptual influence of the experimental laboratory. In this regard, it is somewhat more than coincidental that the historical roots of an extensive research literature in the field can be traced to the early studies of Pavlov and Sherrington who, before the turn of the century, focused attention upon the central role of behavioral interactions in the physiological adaptations and adjustments of the internal environment (Pavlov, 1927, 1928; Sherrington, 1906). Of at least as great import, however, was the foundation provided by these early investigations, as well as by those of Beckterev (1932) in the Soviet Union and Thorndike (1898) in the United States, for conceptualizing the behavioral interactions between organism and environment within the framework of an orderly and systematic body of scientific knowledge based upon observation and experiment.

It is unfortunately true that the lessons learned about behavior as a consequence of such controlled laboratory investigations have not always been warmly embraced at the clinical level for reasons that appear to be somewhat unique to the human condition. Unlike other aspects of biology (anatomy, physiology, biochemistry), where behavior is concerned we bipeds harbor strong chauvinistic dispositions which have little to do with gender, race, or other personal characteristics of the species. Acceptance of experimental behavior analysis with lower organisms as instructive with respect to exalted human performance repertoires has been hard won in the face of vigorous

Joseph V. Brady • Division of Behavioral Biology, The Johns Hopkins University, Baltimore, Maryland 21205.

higher-order resistance. It is nonetheless true that traditional appeals to unobserved and unobservable mental processes and other explanatory fictions must now yield to more operational analyses of clinically-relevant behavioral interactions based upon detailed and objective descriptions of observable and quantifiable events.

But what, in fact, have we learned in the laboratory about the nature of those activities at the interface between individuals and their environments which define the unique domain of behavior analysis? At the most fundamental level, there appear to be two basic modes that characterize this interactive process. In the first instance, a reactive mode is clearly rooted in the biochemical and physiological adaptations of the organism to the influences of a changing environment (i.e., the environment acts upon the organism and the organism reacts). Since at least the time of Pavlov, this respondent paradigm has provided the basis for describing and experimentally analyzing increasingly more complex interactions of direct relevance to clinical medicine, in general, and to cardiovascular adaptations in particular. Early respondent conditioning studies (Dykman and Gantt, 1958; Deane and Zeaman, 1958) provided systematic accounts of how neutral environmental stimuli (e.g., tones and lights), which initially produced only minimal changes in circulatory activity, could elicit conditional cardiovascular responses (e.g., heart rate increases) of substantial magnitude and duration when paired repeatedly with unconditional environmental stimulus events (e.g., food or electric shock) that normally elicited such changes. If such conditional tone or light stimuli (i.e., CS) are subsequently presented a number of times without the unconditional food or shock stimuli (i.e., UCS), the magnitude and frequency of the conditional heart rate increase response (i.e., CR) elicited by the CS diminish, and respondent extinction occurs. When a period of time intervenes between such extinction and subsequent presentations of the CS, however, spontaneous recovery of the CR is observed in the form of temporary reappearance of the response elicited by the CS.

The power to elicit a CR, which is developed in one CS by conditioning, extends to other stimuli, with the degree of this "stimulus generalization" determined by the similarities and differences between the other stimuli and the CS. Because stimuli other than the CS differ with respect to the magnitude and frequency with which they elicit the CR, "stimulus discrimination" also occurs. Indeed, discrimination can be made increasingly more pronounced by repeated pairings of the UCS only with a specific CS (i.e., respondent conditioning) while insuring that the occurrence of other stimuli is not paired with the UCS.

These basic observations with regard to the reactive or respondent conditioning mode have been elaborated in numerous laboratory and clinical–experimental studies since Russian researchers first introduced this systematic approach to behavior analysis. It has been convincingly demonstrated, for example, that second- or higher-order conditioning can occur when a well-established CS is paired with a neutral stimulus. The neutral stimulus acquires the power to elicit the reactive CR. Although it has not been empirically determined just how far this process can be carried, the development of eliciting properties by CSs two or three steps removed from the original UCS is not uncommon. And the intensive investigative effort, principally Russian in origin, to extend the conceptual framework of such classical or Pavlovian (i.e., respondent) conditioning to encompass verbal stimuli and somatic responses (Razran, 1961) suggests potentially important directions for development of the theory and the practice of behavioral medicine.

Elicited responses of the type that have provided the primary focus for such basic and important respondent or reflex conditioning analyses must none-theless be seen to represent only a relatively small proportion of the behavioral interactions of higher organisms. The most prominent aspects of such advanced repertoires are represented by the second basic, and generally more active than reactive, mode characterizing behavioral interactions focusing on the opera-tions performed by organisms upon their internal and external environment rather than upon their "reflex" reactions to such environmental influences.

The frequency of such actively operant behavior is chiefly determined by the environmental consequences of that behavior. When these environmental consequences increase the likelihood that the behavior will recur, "reinforce-ment" is defined. When, on the other hand, the consequences of an operant performance decreases the likelihood of that behavior recurring, "punishment" is defined. The important point to be made here is simply that reinforcement and punishment are always defined by the effects of these operations on the frequency or strength of behavioral interactions.

Over the past three decades, a broad range of animal laboratory and human experimental studies has provided important insights into the principles that determine the acquisition, maintenance, and modification of such operant behavior (Honig, 1966; Honig and Staddon, 1976). The basic observation is that the rate of an operant response already in the organism's repertoire can be readily increased by reinforcement (operant conditioning). Beyond this, it has been possible to make explicit the process called "shaping", whereby oper-ant conditioning can extend existing simple responses into new and more com-plex performances. Of critical importance for this shaping process is the obser-vation that a reinforcer not only strengthens the particular response that precedes it, but also produces an increase in the frequency of many other sim-ilar bits of behavior, and, in effect, raises the individual's general activity level.

Thus, the shaping of behavior proceeds as reinforcers are initially pre-sented following a response similar to or approximating the desired one. Since this tends to increase the strength of various other similar behaviors, a response still closer to the desired one can be selected from this new array and can be reinforced. Continued narrowing and refinement of the response criteria required for reinforcement leads progressively to new arrays of available behavior. In this way, by successive and progressive approximation, a new and desired performance can be shaped. The importance of this simple but fundamental and powerful shaping process for the development and modification of behavior can not be overstated, since the weight of available evidence suggests that a careful and systematic application of such procedures with effective reinforcers is sufficient to establish or alter any operant performance of which the organism is physically capable. This shaping process is obviously of enormous clinical importance in behavioral medicine since many patient performances can only effectively be changed in this way. Without shaping, one might wait for inordinately long periods before a patient performs some critical health-related behavior that could be reinforced.

The fact that changes in behavior are not always brought about by deliberate and systematic manipulation of the environment, however, has led to an analysis of superstitious behavior. A potentially reinforcing environmental event may, by chance, follow a response, resulting in the adventitious strengthening of that response. If this sequence of events recurs even infrequently (i.e., intermittent reinforcement, as described below), the individual may learn quite elaborate sequences of superstitious behavior which have absolutely nothing to do with production of the event that is influencing the frequency of the behavior. The elaborate rituals of the gambler do not produce winning dice combinations any more than native dances produce rain; they persist because they are occasionally followed by "7 or 11," in the first instance, and precipitation in the second.

The powerful effects of reinforcement in establishing and maintaining operant behavior suggest that withholding such reinforcing consequences (i.e., extinction) will have comparably powerful effects on the strength of previously reinforced responses. Indeed, such extinction procedures do reduce the frequency of response, although the reduction is not usually immediate. Rather, after the onset of extinction, the initial effect is often a brief increase in the frequency as well as the force and variability of the response previously followed by reinforcement. The extent to which operant responding persists in the absence of reinforcing environmental consequences (i.e., resistance to extinction) depends, of course, on the interaction of many complex influences including motivational factors (e.g., level of deprivation). But both laboratory and clinical experimental evidence now confirms that the single most important variable affecting the course of operant extinction is the schedule of reinforcement on which the performance was previously acquired and maintained.

Whenever a reinforcing environmental stimulus follows some but not all occurrences of an operant response, a schedule of intermittent reinforcement is operating. Accordingly, then, intermittent reinforcement is defined when only selected occurrences of an operant are followed by a reinforcer. Every reinforcer occurs according to some schedule or rule, although some schedules are so complicated that detailed analysis is required to formulate them precisely. Simple schedules of intermittent reinforcement can be classified into two broad categories: ratio and interval schedules. Ratio schedules prescribe that a certain number of responses be emitted before one response is reinforced, the term "ratio" referring to the relationship between the required response total (e.g., 50) and the one response followed by the reinforcing event (e.g., piecework schedule requiring 49 discrete labor units before the single 50th performance is followed by payoff). Interval schedules on the other hand prescribe that a given interval of time elapse before an emitted response can be followed by a reinforcing event. The relevant interval can be measured from any event, but the occurrence of a previous reinforcer is usually used (e.g., salaried pay schedules). The recuperative properties of interval schedules under which the mere passage of even long time intervals brings an opportunity for a single response to be followed by a reinforcer contrasts with the strain potential of high ratio schedule requirements under which the performance may extinguish before a sufficient number of responses are emitted for one to be followed by reinforcement (Rachlin, 1970).

Even simple ratio and interval schedules can in turn be classified into two general categories based upon whether the required number of responses or lapse of time are fixed or variable, and all known schedules of reinforcement can be reduced to variations of these basic ratio and interval parameters. A single operant performance, for example, may be followed by a reinforcer in accordance with the requirements of two or more schedules at the same time, or compound schedules. Two or more responses may be followed by a reinforcer according to the requirements of two or more schedules at the same time, or concurrent schedules. And perhaps the most ubiquitous case of reinforcement schedule complexity is represented by the multiple schedule, under the requirements of which two or more independent schedules developed and maintained simultaneously are called forth sequentially under discriminably different environmental stimulus conditions. Virtually all operant behavior is followed by reinforcing stimuli according to multiple, compound, and concurrent schedules built out of the same basic elements as the simple ratio and interval schedules. Each schedule, simple or complex, generates and maintains its own characteristic performance, and when reinforcement is discontinued, the course and character of extinction are prominently influenced by the preceding schedule of reinforcement. Significantly, it has also become increasingly clear in the laboratory and the clinic that at least the frequency or rate of a given operant performance can be more effectively controlled by reinforcement schedule manipulation than by any other means.

The detailed experimental analysis of reinforcement schedules has also served to emphasize another very important set of relationships between operant performances and environmental events encompassed within the general conceptual framework of stimulus control. The occurrence of a reinforcer following an operant not only increases the likelihood that the response will recur but it also contributes to bring that performance under the control of other environmental stimuli present when the operant is reinforced. After the responses composing the operant have been reinforced in the presence of a particular stimulus a number of times, that stimulus comes to control the operant (i.e., the frequency of those responses is high in the presence of the stimulus and lower in its absence). A discriminative stimulus is thus defined by this process as one in whose presence a particular operant performance is highly probable because the behavior has previously been reinforced in its presence. It is important to recognize, however, that discriminative stimuli do not elicit performances as in the respondent or reflex case, but rather set the occasion for operant responses in the sense that they provide the circumstances under which the performance has previously been reinforced. The control over driving behaviors by traffic signals occasioning vehicle braking and accelerating occurs because of systematic relationships between such performances and their consequences (e.g., fines and accidents), not because of any inherent or conditional eliciting properties of red, green, and yellow lights. This controlling power of a discriminative stimulus develops gradually, and at least several occurrences of the reinforcer following the response in the presence of the stimulus are required before the stimulus effectively controls the performance.

Such discriminative stimulus control is not an entirely selective process, however, since reinforcement of a performance in the presence of one stimulus increases the tendency to respond not only in the presence of that stimulus but also in the presence of other stimuli with similar properties. This is stimulus generalization. It is not always clear from simple observation, of course, which stimulus or which property of a stimulus is controlling an operant performance, and both laboratory and clinical experiences have documented the hazard of assuming that the similarity casually observed between stimuli provides an adequate explanation of such generalization. There is unfortunately no substitute for experiment in differentiating the many detailed aspects of a stimulus complex that may exercise critical control. Furthermore, related response generalization effects have also been observed to occur when following an operant with a reinforcer results not only in an increase in the frequency of the responses composing that operant but also in an increase in the frequency of similar responses.

This very sensitivity to the differential aspects of stimulus and response complexes provides the basis for the other major cornerstone of the stimulus control process identified as discrimination. A discrimination between two stimuli is said to obtain when an organism behaves differently in the presence of each. Such stimulus discrimination is pronounced under conditions which provide differential reinforcement. This process is seen to operate in the formation of a discrimination when there is a high probability that a reinforcer will follow a given response in the presence of one stimulus, and a low or zero probability that reinforcement will follow the response in the presence of another stimulus. The extent of the generalization between two stimuli will influence the rapidity and stability with which a discrimination can be formed, and it is important to recognize that the antecedents of a performance that occurs under one set of stimulus conditions may include events which have occurred under quite different stimulus conditions. The careful application of differential reinforcement procedures can, nonetheless, bring about remarkably precise control of an operant performance by highly selective aspects of a stimulus complex. This attention to specific properties of a stimulus can be facilitated and enhanced by the use of instructional stimuli, which tell about features of the environment that are currently relevant to the occasioning of reinforcement. An example is a treasure map. Imitation and modeling, considered analytically, appear to represent special case instances of such instructional control. Furthermore, this precise stimulus control (i.e., "attention") can be transferred from one group of stimuli or stimulus properties to another by simultaneous presentation of the two followed by the gradual withdrawal or fading of the original stimulus.

The intimate and continuing association between discriminative environmental stimulus events and the occurrence of reinforcement endows at least some originally nonreinforcing stimuli with acquired reinforcing properties. These stimuli, fraternity pins and stock market quotations, for example, have come to be designated as secondary or conditioned reinforcers to distinguish them from innate, primary, or unconditioned reinforcers which require no experience to be effective. Such conditioned reinforcers can be either appetitive (i.e., strengthening prior-occurring responses by their appearance) or aversive, in which case their removal or postponement is reinforcing. The development or acquisition of conditioned reinforcing properties by a stimulus is usually a gradual process, as is the case with discriminative stimuli in general. A common interpretive view of the process suggests that conditioned reinforcers may owe their effectiveness to the fact that they function as discriminative stimuli for later members of a response chain which are maintained by the occurrence of reinforcers in their presence.

Response chaining refers to the observationally and experimentally verified occurrence of a composed series of performances joined together by environmental stimuli that act both as conditioned reinforcers and as discriminative stimuli. A chain (e.g., party-going) usually begins with the occurrence of a discriminative stimulus (e.g., phone invitation) in the presence of which an appropriate response (e.g., acceptance) is followed by a conditioned reinforcer (e.g., "Glad you can make it."). This conditioned reinforcer is also the discriminative stimulus occasion for succeeding response (e.g., bathing, shaving, and dressing), which in turn is followed by another conditioned reinforcer (e.g., leaving the house, catching a cab), which is also a discriminative stimulus for the next response (e.g., joining the party), and so on. While it is doubtless true that the entirety of such chains is most often maintained by the terminal occurrence of potent environmental consequences (e.g., social interactions, food, sex), laboratory experiments have clearly demonstrated that the overlapping links in the chain (i.e., discriminative stimulus → operant response → conditioned reinforcer) are held together primarily by the dual (and demonstrably separable) discriminative and conditioned reinforcing functions of environmental stimuli. The significance of this general chaining principle must be seen to reside in the fact that virtually all behavioral interactions occur as chains of greater or lesser length, and that even performances usually treated as unitary phenomena can be usefully analyzed at various component levels (e.g., golf, bowling, tennis) for purposes of modification or proficiency enhancement.

Perhaps the most important aspect of this complex analysis of environmental stimulus events in relation to behavioral interactions is the clear implication that some degree of independence can be gained from the factors limiting conditioned-reinforcer potency by the formation of conditioned reinforcers based upon two or more primary reinforcers. Such conditioned stimulus events (generalized reinforcers) gain potency from all the reinforcers on which they are based, and verbal behavior, the most prominent operant performance in the human repertoire, as well as the most valued stimulus consequence in the social environment (money) can be seen to share these broadly-based discriminative and generalized conditioned reinforcing properties.

This necessarily abbreviated overview of experimentally-derived concepts and principles relevant to medicine in general and cardiovascular risk factors in particular has thus far maintained the traditionally accepted differentiation between active (operant) and reactive (respondent) behavioral interaction modes based principally upon procedural distinctions identified in the laboratory. The independent and distinctive features of these two coextensive processes are seldom apparent, however, in the course of even detailed natural observation. In no investigative aspect of the behavioral universe is the complex interaction between these active and reactive modes more pronounced than in the experimental analysis of aversive control procedures represented (or *misrepresented*!) by the technical terms punishment, escape, avoidance, and their emotional and motivational corollaries.

Empirical and theoretical accounts of those aspects of behavioral medicine concerned with disordered performances frequently assign a central role to historical and contemporary environmental interactions involving aversive circumstances and conditions. Operationally characterized in terms of their behavioral effects, aversive stimuli are defined as environmental events which decrease the subsequent frequency of the operant responses they follow, on the one hand, and/or increase the subsequent frequency of operant responses which remove or postpone them. When an aversive stimulus follows an operant, and decreases the likelihood that such performances will recur, a punishment condition is defined. Punishment may be made contingent upon the occurrence of an operant which has never before been followed by a reinforcer, an operant currently being maintained by appetitive or aversive reinforcement, or an operant that is undergoing extinction. Under each condition, the short- and long-term effects of punishment will vary as a function of complex operant-respondent interactions, and both discriminative stimulus control and reinforcement schedule factors may operate to further influence the subsequent form and frequency of the performance.

An escape condition is defined when a response terminates an aversive stimulus after the stimulus has appeared. The interaction between operants and respondents is especially prominent in escape situations, since the aversive stimulus usually elicits reflexive responses which eventually result in or accompany an operant performance followed by withdrawal of the aversive stimulus. Strong generalization effects appear during initial exposures to escape situations, but the gradual development of discriminative properties by the aversive stimulus narrows the performance, and very low intensities of the aversive stimulus may eventually maintain an operant escape performance requiring a much more intensive aversive stimulus to establish. Reinforcement schedule effects similar in all essential respects to the appetitive conditions described above are observed when withdrawal of an aversive stimulus is the reinforcer. Extinction of an operant escape response occurs rapidly when presentation of the aversive stimulus is discontinued, or more slowly and erratically if the occurrence of the operant is no longer reinforced by withdrawal of the recurring aversive stimulus.

An "avoidance" condition is defined by the occurrence of an operant response that postpones an aversive stimulus. Avoidance performances may be established and maintained either in the presence or absence of an exteroceptive environmental event (i.e., warning stimulus) which precedes the aversive stimulus. When an exteroceptive warning stimulus precedes the aversive stimulus, respondent conditioning effects operate to endow the warning stimulus with aversive properties, the termination of which following the operant avoidance response probably combines with the continued absence of the aversive stimulus to act as a reinforcer. The complexity of the avoidance process is suggested by the functionally simultaneous properties acquired by the conditioned aversive warning stimulus as 1) an eliciting environmental event for respondent behaviors, 2) a conditioned aversive reinforcer, withdrawal of which strengthens the operant avoidance performance effective in removing it, and 3) a discriminative stimulus which provides the occasion for the operant avoidance response to be followed by a reinforcer. In the absence of an exteroceptive warning stimulus, a temporal respondent conditioning process provides discriminative cues, and the temporal stimulus correlated with the aversive environmental event acquires the same three simultaneous functions as an exteroceptive stimulus.

Such an analysis of aversive control emphasizes the simultaneous operation of active or operant and reactive or respondent conditioning processes in ongoing behavior segments. Whenever the conditioned stimulus in a respondent conditioning procedure is an appetitive or aversive reinforcer, operant conditioning occurs at the same time as respondent conditioning. Similarly, whenever the reinforcer in an operant procedure is an unconditioned stimulus, respondent conditioning proceeds at the same time as operant conditioning. Thus, insofar as the eliciting and reinforcing stimulus classes are composed of the same environmental events, operant and respondent processes are coextensive.

Relevant applications of these basic behavioral principles to clinical medicine in general and to the analysis of cardiovascular risk factors in particular have emerged in two major forms. In the first instance, procedures have been developed for active rather than reactive behavioral control of visceral, somatomotor, and central nervous system processes based upon the arrangement of explicit contingency relationships between specific antecedent physiological events and programmed environmental consequences. It has been convincingly demonstrated that such behavioral "biofeedback" intervention can produce reliable bidirectional control over both increases and decreases in cardiac rate (DiCara and Miller, 1969; Engel and Gottlieb, 1970) and blood pressure (Pappas, DiCara, and Miller, 1970; Benson, Herd, Morse, and Kelleher, 1969). Large magnitude and enduring elevations in heart rate (Harris, Gilliam, and Brady, 1976) and blood pressure (Harris, Findley, and Brady, 1971) have also been described in more chronic operant conditioning studies. Significantly, more recent studies (Harris, Gilliam, Findley, and Brady, 1973) have involved the application of operant shaping techniques with both amplitude and duration of blood pressure elevations systematically increased in small progressive steps to diastolic pressures 35 to 40 mm/Hg above pre-experimental resting levels.

Figure 1 shows the relative frequency distributions of diastolic blood pressure from an experiment in which a baboon learned to increase and maintain blood pressure elevations in order to obtain food and avoid shock (Turkkan and Harris, 1980). The shaping procedure involved delivery of food pellets for accumulation of 600 sec of time above the diastolic pressure criterion level, and delivery of a single electric shock to the tail for accumulation of 240 sec of time

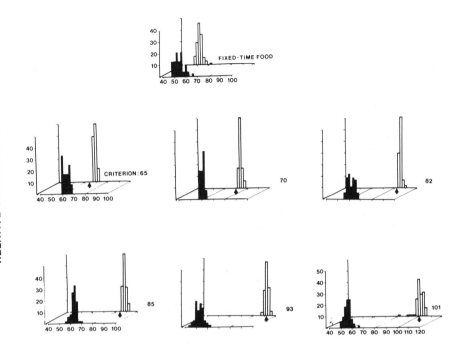

DIASTOLIC BLOOD PRESSURE (mmHg)

Figure 1.   Relative frequency distributions of 40-minute average diastolic pressures for baboon "#82" during a baseline condition (fixed-line food), and at successively higher diastolic criteria (columns, from right to left). Open bars represent diastolic pressure levels from 4 experimental sessions, while filled bars represent data from 4 associated postsession periods. Arrows indicate the diastolic criterion level at each stage of training, with criterion values shown numerically to the right of each graph.

below that criterion level. When the pressure level was above criterion, white light appeared on the animal's work panel, and when pressure was below criterion, a red light accompanied by 1000 Hz tone was presented. Experimental sessions began at noon each day, and ended at midnight. Criterion levels beginning at 65 mm Hz (i.e., pre-experimental baseline average diastolic pressure level) were progressively elevated at a rate approximating 2–3 mm/Hg per week. The systematic shaping of diastolic pressure elevations over a 10–12 week conditioning period, illustrated in Figure 1, compares the diastolic pressure levels recorded *during sessions (open bars)* with the levels recorded during the 12-hour intervals *between sessions (filled bars)* under baseline conditions (top segment) and during successive stages of conditioning. At the highest criterion (lower right segment), diastolic pressures were elevated above 100 mm of Hg in order to maintain a food-abundant environment throughout the 12-hour experimental session, during which less than one shock per hour was delivered. Remarkably, there was no overlap between the distributions of pressure levels recorded at this highest criterion and those recorded during the baseline period.

While these observations clearly reflect the participation of an active behavioral process in the development and maintenance of cardiovascular functions traditionally considered under more reactive control, there is implied no claim to exclusivity. Multiple mechanisms, both behavioral and physiological, must be presumed operative in the mediation of such complex psychophysiological interactions. Insofar as the internal and external environmental stimulus events involved in these processes have common functional properties (e.g., eliciting, reinforcing, discriminative), both operant and respondent conditioning, at the very least, can be considered coextensive.

The second significant development that has emerged in the context of these behavior analysis principles involves the application of contingency management procedures for the shaping, maintenance, and modification of health-related performances (e.g., food intake, exercise, and medication compliance) in the interest of cardiovascular risk reduction (e.g., Squyres and Coates, 1980). Of particular relevance in this regard would seem to be the experimentally and clinically documented effects of scheduling conditions, stimulus control, and chaining which determine under what circumstances and in accordance with what behavioral requirements a valued commodity service or substance can be obtained. The demonstrably potent influence of these factors upon the strength and persistence of behavior is worth emphasizing because these properties of a performance frequently appear as the most baffling and recalcitrant aspects of risk-factor reduction (e.g., smoking and overeating). Indeed, it is to the power of the kind of environmental constraints imposed by scheduling and stimulus control procedures to entrain performances of remarkable persistence that particular attention would seem appropriately directed.

Figure 2 illustrates a typical segment of a cumulative record from an experiment in which a chimpanzee sustained performance on a ratio-schedule which required 120,000 responses on a heavy push-button manipulandum for access to food (Findley and Brady, 1965). After each 4,000 responses toward the total requirement, a brief flash of light was presented—the same light that was illuminated continuously during food access once the total ratio was completed. Of particular interest is the pause which follows each flash of light after a block of 4,000 responses, illustrating the control acquired by this conditioned-reinforcing stimulus event. Subsequent extension of a 250,000 response ratio and manipulations involving removal and reintroduction of the light flash after each 10,000 responses documented the critical interactions between rule-gov-

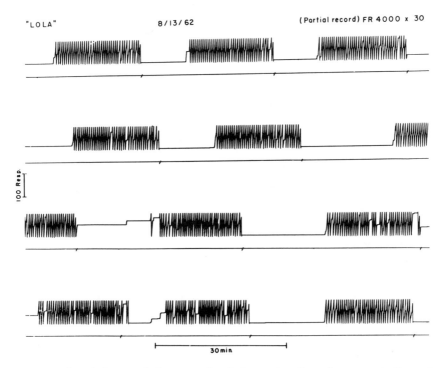

Figure 2. Cumulative record of responses (vertical excursion of stepping pen-reset after each 100 responses) over time (horizontal baseline-paper speed) for chimpanzee "Lola" showing pause following each flash of light (conditioned reinforcer) after a block of 4,000 responses toward the total requirement of 120,000 for access to food.

ernance and stimulus control in the establishment and maintenance of such remarkably persistent performance repertoires. It seems important to recognize that while such unusual and extreme examples of schedule and stimulus conditions may appear to push the limits of adaptive functions, they are not tricks or circus acts. They do in fact represent the orderly and lawful operation of general relationships common to all behavioral interactions, including cardiovascular risk factors, and appear to be of particular relevance to the excessive or abusive aspects of such performances.

Acknowledgments

Research reported in this manuscript supported in part by NIDA Grant DA00018, NHLBI Grants HL17958, HI17970, and NIMH Grant MH15330.

References

Bechterev, V. *General principles of reflexology*. Translation: E. In Murphy and W. Murphy (Eds.), London: Hutchinson, 1932.

Benson, H., Herd, J. A., Morse, W. H., and Kelleher, R. T. Behavioral inductions of arterial hypertension and its reversal. *American Journal of Physiology*, 1969, *217*, 30–34.

Deane, G. E., and Zeaman, D. Human heart rate during anxiety. *Perceptual and Motor Skills*, 1958, *8*, 103–106.

DiCara, L. V., and Miller, N. E. Transfer of instrumentally learned heart rate changes from curarized to noncurarized state: Implications for a mediational hypothesis. *Journal of Comparative and Physiological Psychology*, 1969, *62* (2, Pt. 1), 159–162.

Dykman, R. A., and Gantt, W. H. Cardiovascular conditioning in dogs and in humans. In W. H. Gantt (Ed.), *Physiological bases of psychiatry*. Springfield, Ill.: C. C. Thomas, 1958.

Engel, B. T., and Gottlieb, S. H. Differential operant conditioning of heart rate in the restrained monkey. *Journal of Comparative and Physiological Psychology*, 1970, *73* (2), 217–225.

Findley, J. D., and Brady, J. V. Facilitation of large ratio performance by use of conditioned reinforcement. *Journal of Experimental Analysis of Behavior*, 1965, *8*, 125–129.

Harris, A. H., Findley, J. D., and Brady, J. V. Instrumental conditioning of blood pressure elevations in the baboon. *Conditional Reflex*, 1971, *6* (4), 215–226.

Harris, A. H., Gilliam, W. J., Findley, J. D., and Brady, J. V. Instrumental conditioning of large magnitude daily 12-hour blood pressure elevations in the baboon. *Science*, 1973, *183*, 175.

Harris, A. H., Gilliam, W. J., and Brady, J. V. Operant conditioning of large magnitude 12-hour heart rate elevations in the baboon. *Pavlovian Journal of Biological Sciences*, 1976, *11* (2), 86–92.

Honig, W. K. (Ed.). *Operant behavior: Areas of research and application*. New York: Appleton-Century-Crofts, 1966.

Honig, W. K., and Staddon, J. E. R. (Eds.). *Handbook of operant behavior*. Englewood Cliffs, New Jersey: Prentice Hall, Inc., 1976.

Pappas, B. A., DiCara, L. V., and Miller, N. E. Learning of blood pressure responses in the noncurarized rat: Transfer to the curarized state. *Physiology and Behavior*, 1970, *5* (9), 1029–1032.

Pavlov, I. P. *Conditioned reflexes*. (G. V. Anrep., Trans.) London: Oxford University Press, 1927.

Pavlov, I. P. *Lectures on conditioned reflexes*. (W. H. Gantt, Trans.) New York: International Press, 1928.

Rachlin, H. *Introduction to modern behaviorism*. San Francisco, Ca.: W. H. Freeman and Company, 1970.

Razran, G. The observable unconscious and the inferable conscious in current Soviet psychophysiology: Interoception conditioning, semantic conditioning, and the orienting reflex. *Psychological Review*, 1961, *68*, 81–147.

Sherrington, C. S. *The integrative action of the nervous system*. 1947 edn. England: Cambridge University Press, 1906.

Squyres, W. D., and Coates, T. J. A self-management approach to cardiovascular risk reduction: Management of the self and the environment. *Health Education Monographs*, 1980 (in press).

Thorndike, E. L. Animal intelligence—an experimental study of the associative processes in animals. *Psychology Monograph*, 1898, *2*, 1–106 (Mongr. suppl. whole no. 8).

Turkkan, J. S., and Harris, A. H. Differentiation of blood pressure elevations in the baboon using a shaping procedure. *Behavioral Analysis Letters*, 1981, *1*, 97–106.

Chapter 7

Behavioral Processes and Treatment Outcomes in Coronary Heart Disease

Redford B. Williams, Jr.

It has been estimated that in 1979 110,000 coronary artery bypass graft (CABG) operations were performed at an average cost of $15,000 per patient—a total cost of $1.65 billion (Kolata, 1981). While it has been shown in several studies (Conley, Wechsler, Anderson, Oldham, Sabiston and Rosati, 1977; Whalen, Wallace, McNeer, Rosati, and Lee, 1977; Kloster, Kremkau, Ritzmann, Rahimtoola, Rosch, and Kanarek, 1979; Hammermeister, DeRouen and Dodge, 1979; Harris, Phil, Harrell, Lee, Behar, and Rosati, 1979) that survival is prolonged with CABG treatment in patients with left-main disease and, possibly, three-vessel disease, patients with these grades of coronary atherosclerosis make up only about 43% of patients in the National Heart, Lung, and Blood Institute's Coronary Artery Surgery Study (Kolata, 1981). Thus, over half the patients currently being treated with CABG surgery cannot logically be expected to survive longer as a result of the surgery than patients receiving medical management alone. However, what does appear clear from the cited studies, as well as from the recent concensus conference on CABG surgery (Kolata, 1981), is that even among patients with one- and two-vessel disease relief of anginal pain from surgery is superior to that from medical management alone, and it is to achieve this benefit of angina relief that the majority of CABG operations are performed. If the approximately 62,700 patients per year with one- and two-vessel disease who are currently failing to obtain adequate control of angina with medical management and who go on to CABG could be prospectively identified and given special attention prior to embarking on the eventually unsuccessful trail of medical management, it is conceivable that a substantial reduction in costs of treatment, as

Redford B. Williams • Duke University Medical Center, Department of Psychiatry, Box 3416, Durham, North Carolina 27710.

well as reduction in pain and suffering and risks associated with CABG surgery, could be achieved in a significant proportion.

While various indices of coronary atherosclerosis and left ventricular function are potent predictors of survival among both medically-treated (Conley *et al.*, 1977; Kloster *et al.*, 1979; Hammermeister *et al.*, 1979) and surgically-treated (Conley *et al.*, 1977; Kloster *et al.*, 1979) patient groups, there is no presently known comparable set of physical characteristics that predict pain relief from either medical or surgical management. The question addressed in the remainder of this chapter is this: Are there behavioral or psychosocial characteristics of patients with coronary heart disease (CHD) which do predict important outcomes? If so, assessment of these characteristics could serve to identify patient groups in whom more intensive or new preventive or treatment approaches might be undertaken with the goal of improving outcomes. The chapter by Syme in this volume presents persuasive evidence that persons with low levels of social support are at greater risk of dying from all causes, including CHD, compared to persons with higher levels of social support. Some evidence regarding the relation of behavioral and psychosocial characteristics to two other CHD "outcomes"—arteriographically documented coronary atherosclerosis and relief of anginal pain—shall be reviewed. With regard to the former, prospective identification of predisposing behavioral and psychosocial factors might point the way to preventive measures that could reduce the rate of atherogenesis. Much the same rationale underlies the current efforts to achieve dietary changes with the goal of reducing the contribution of hyperlipidemia to atherogenesis. With regard to pain relief, prospective identification of psychosocial and behavioral factors that predispose patients to fail to obtain relief of angina could help to single out patient groups among whom special attention in terms of more intensive pharmacologic treatment or innovative behavioral treatment approaches could serve to improve the pain relief outcome.

The chapter by Glass in this volume presents the evidence supporting the association between Type A behavior pattern and increased rates of CHD events, an association which has now gained wide acceptance (Review Panel on Coronary-Prone Behavior and Coronary Heart Disease, 1981). In addition to this association with clinical events, there is now extensive evidence suggesting that arteriographically-documented coronary atherosclerosis is more severe and extensive among persons displaying the Type A behavior pattern. Four studies existing have reported a positive relationship between more severe and extensive coronary atherosclerosis and Type A behavior pattern as assessed by the structured interview technique (Blumenthal, Williams, Kong, Schanberg, and Thompson, 1978; Frank, Heller, Kornfeld, Sporn, and Weiss, 1978; Krantz, Sanmario, Selvester, and Matthews, 1979) or the Jenkins Activity Survey (JAS) (Zyzanski, Jenkins, Ryan, Flessas, and Everist, 1976). Two other studies from Dimsdale's group in Boston have failed to find a relationship between coronary atherosclerosis and Type A behavior pattern assessed by either the structured interview (Dimsdale, Hackett, Hutter, Block, and Catanzano, 1979) or the JAS (Dimsdale, Hackett, Hutter, Block, and Catanzano, 1978). It is unclear at present whether the failure of Dimsdale *et al.* (1978, 1979) to find a positive relationship between coronary atherosclerosis and Type A behavior pattern is a function of some as yet undefined differences between their methods and/or patient population and those of the four studies reporting a positive association (Blumenthal *et al.*, 1978; Frank *et al.*, 1978; Krantz, *et al.*, 1979; Zyzanski *et al.*, 1976). Nevertheless, the weight of evidence currently available suggests the existence of a positive relationship between Type A behavior pattern and coronary atherosclerosis, at least among the population of patients referred for diagnostic coronary arteriography. This suggests, in turn, that the Type A behavior pattern is involved in the process of atherogenesis as well as in the precipitation of acute clinical CHD events. If Type A behavior does indeed play some role in coronary atherogenesis, primary prevention programs aimed at reducing atherogenesis rates should address the issue of modifying the Type A behavior pattern, as well as such other known risk factors as cigarette smoking, high blood pressure, and hyperlipidemia. Recent research at Duke regarding the relation between behavior and arteriographically-documented coronary atherosclerosis has shown that Type A behavior as well as the subcomponent of hostility, is associated with more severe coronary atherosclerosis.

Figure 1 shows the relationship between Type A behavior pattern and severity of coronary atherosclerosis (Blumenthal *et al.*, 1978). TOTCI scores

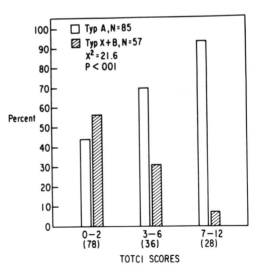

Figure 1. Proportions of Type A and non-Type A (X or B) patients among groups of patients with mild (TOTCI score = 0–2), moderate (TOTCI score = 3–6) and severe (TOTCI score = 7–12) coronary atherosclerosis on arteriography. (Taken from Blumenthal *et al.*, 1978.)

are a measure of extent and severity of coronary atherosclerosis based on the degree to which each of the four major coronary arteries is occluded. Among patients with only mild coronary atherosclerosis (TOTCI <3), Type A and Type B patients are present in about equal proportions. With moderate (TOTCI = 3–6) and severe (TOTCI >6) atherosclerosis, however, the proportion of Type A patients increases to 70% and 93%, respectively. The statistical significance of this increasing proportion of Type A patients in groups with increasingly extensive and severe coronary atherosclerosis is not diminished when age, sex, blood pressure, serum cholesterol, and history of cigarette smoking are all simultaneously adjusted for in an analysis of covariance (Blumenthal et al., 1978). Thus, Type A behavior pattern is associated with an increase in extent and severity of arteriographically-documented coronary atherosclerosis, and this relationship is statistically independent of any influence of traditional risk factors. Moreover, since it is independent of sex, it also appears that Type A behavior pattern may predispose to more severe coronary atherosclerosis among women as well as men.

Subsequent to our initial study described above, we have reported further on the joint relationship between Type A behavior pattern and hostility (assessed using a subscale from the MMPI) and coronary atherosclerosis (Williams, Haney, Lee, Kong, Blumenthal, and Whalen, 1980). Of 424 patients undergoing coronary arteriography at Duke University Medical Center, 71% of the Type A patients had a clinically significant occlusion (defined as 75% or greater diminution of luminal diameter) in at least one coronary artery. In contrast, only 56% of the non-Type A patients showed evidence of a clinically significant occlusion. In addition, scores (Ho) on the Cook and Medley (1954) Hostility scale were related to presence of a clinically significant coronary occlusion: Patients with Ho scores of 10 or less out of a possible maximum of 50 showed only a 48% rate of clinically significant occlusions on the coronary arteriogram. In contrast, patient groups scoring higher than 10 showed a 70% rate of significant occlusions. Figure 2 shows the joint relationship between Ho score and Type A behavior pattern and the presence of clinically significant coronary atherosclerosis, with control for sex differences in Ho score and Type A behavior pattern. It can be seen that both Type A behavior pattern and Ho score (broken down between those scoring 10 or less and those scoring more than 10) are independently related to presence of clinically significant coronary atherosclerosis. Among non-Type A women scoring 10 or less on the Ho scale, only a 12.5% rate of clinically significant coronary atherosclerosis is observed. In contrast, among Type A women scoring higher than 10 on the Ho scale, the rate of significant disease is 46%, nearly four times as much. Among Type A men scoring 10 or more on the Ho scale, the rate of significant disease climbs to 83%.

By scoring the Ho scale on MMPI protocols that were completed several

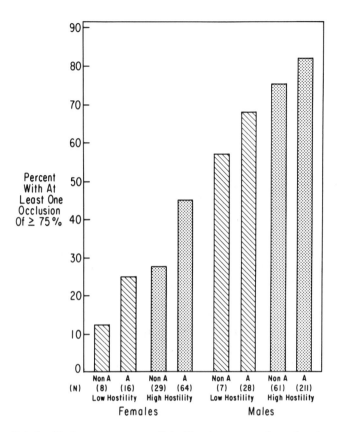

Figure 2. Relationship between presence of significant coronary atherosclerosis on arteriography and sex, Hostility (indexed by the MMPI Hostility subscale), and Type A behavior pattern tassessed using the structured interview). (Taken from Williams *et al.*, 1980.)

years ago in a number of prospective studies that are currently in progress, it will be possible to determine whether MMPI scales that correlate with coronary atherosclerosis—such as the Ho scale—are also predictive of CHD mortality. Very preliminary analysis of data from one such study suggests that persons with Ho scores of 10 or less have a significantly lower CHD mortality compared to persons scoring higher than 10 on the Ho scale. We are currently performing item and factor analyses in over 2,000 patients to identify correlated groups of MMPI items that correlate with coronary atherosclerosis severity. Such item clusters could form the basis of new coronary atherosclerosis scales whose generalizability to the population-at-large could be checked through similar analyses on existing MMPI data sets such as that described above for the Ho scale. Successful completion of such analyses could define psychological traits that prospectively identify persons at risk to develop coronary atherosclerosis as well as the clinical sequelae of this pathological lesion. If such identification were possible, preventive measures might be developed for such persons with the goal of reducing their subsequent risk of manifesting CHD events.

As noted earlier, for the majority of patients undergoing CABG surgery in the United States at the current time, the only documented clinical benefit of surgery over medical management is improved control of anginal pain. Under current standards of cardiological practice, the 57% of patients currently operated on annually with one- and two-vessel disease are allocated to surgery mainly because their angina failed to respond adequately to a trial of medical management. As noted, there are no known physical characteristics of such patients which prospectively identify them as being at risk not to achieve adequate angina relief with medical management. Based on studies showing that patients with low back pain and high scores on the MMPI hysteria (Hy) and hypochondriasis (Hs) scales fail to obtain pain relief with back surgery (Sternbach, 1974; Wiltse and Rocchio, 1975), it was predicted several years ago that high Hy and Hs scores would also identify patients with CHD who are at risk not to obtain relief of anginal pain with medical and surgical management. Recent reported findings (Williams, Haney, Lee, Harrell, McKinnis, Blumenthal, Rosati, King, and Sabiston, 1980) confirm this prediction. There were no physical characteristics, either from the clinical history or from the cardiac catheterization data, that reliably and definitively predicted pain relief in either surgical or medically managed patients. In contrast, both Hy and Hs scores were strongly predictive of pain relief in both treatment groups: the higher the Hy or Hs score, the less likely were patients to achieve pain relief (defined as two-class improvement—NYHA angina classification—at six-months follow-up compared to assessment at the time of coronary arteriography) with either surgical or medical management. In the medical management group, among those with Hy scores less than 70, for example, two-class improvement in angina was observed in nearly 80% of patients. Among those with Hy scores above 80, however, the proportion of medically managed patients with two-class improvement at six-months fell to only 40–50%. In addition, it was found that certain social variables also predicted response of angina to medical management. Among those patients who were employed at the time of initial cardiac catheterization, nearly 70% had obtained two-class improvement by six-months follow-up. In contrast, only 42% of those not working at time of cardiac catheterization had obtained two-class improvement by six-months follow-up.

Based on these findings, it appears that prospective assessment of both psychological (Hy, Hs) and social (work status) characteristics among patients being considered for medical management of their angina could serve to pinpoint those patients most unlikely to respond to medical management with adequate relief. Such patients might be targeted for special attention—e.g., more intensive pharmacological therapy or behavioral therapy such as biofeedback and stress management training—with the prospect that a greater proportion might obtain angina relief. If so, a substantial proportion of those patients who are medical treatment failures and are now allocated to surgical management could be helped to obtain relief, thus avoiding surgery with all of its costs and risks.

How might findings such as those described above be incorporated into current cardiological practice with the goal of improving treatment selection for patients presenting for diagnostic studies? First of all, the finding that fewer than 15% of women who are not Type A and who score 10 or less on the Ho scale of the MMPI have significant coronary occlusions might serve to raise the threshold for performing coronary arteriography in this patient group. Further, more precise delineation of those psychological characteristics that predispose to coronary atherosclerosis, as derived from the MMPI analyses described above, could also serve to help select patients for coronary arteriographic studies on a more rational basis than is now possible, particularly in those cases where preliminary studies, such as treadmill testing, are equivocal.

While the clinical application of psychosocial assessment techniques to improve our capacity to identify patients with significant coronary atherosclerosis will likely require extensive further validation research before such assessment can be incorporated into routine clinical practice, the clinical application of the findings relating to relief of anginal pain appears much closer to routine use. This is because the prediction by MMPI Hy and Hs scales of clinical response of another form of chronic pain—low back pain—has already been established to the extent that MMPI assessment is now employed in routine orthopedic practice in many centers. Furthermore, the strong prediction in our studies of anginal pain relief by these MMPI scales, even when physical characteristics fail to predict, provides strong evidence for the specific utility of MMPI assessment in patients with anginal pain.

While precise data are not available, it is likely that well over 300,000 patients undergo coronary arteriographic evaluation annually in the United States at the present time (this estimate is based on a ratio of about one-third of patients who undergo coronary arteriography eventually going on to CABG surgery). If the routine evaluation of this very large patient population were modified now to include psychosocial assessment—a minimum battery would include Type A assessment, MMPI assessment, and assessment of social situation (e.g., work status and social supports)—the immediate payoff would very likely include better selection of patients for special attention with medical management in whom such special attention could result in a larger proportion achieving satisfactory control of angina. The longer-range payoff could include better definition of those psychological and behavioral characteristics that play a role in atherogenesis, with the result that better selection could be made of patients for arteriographic studies and, eventually, preventive measures developed that could reduce the atherogenetic process in predisposed persons. Such an approach has a solid basis in the analogous current efforts to reduce CHD through reduction of such other risk factors as high blood pressure, hyperlipidemia, and cigarette smoking.

References

Aortocoronary-Artery-Bypass assessment after 13 years. *Journal of the American Medical Association,* 1978, *240,* 1353–1354.

Blumenthal, J. A., Williams, R. B., Kong, Y., Schanberg, S. M., and Thompson, L. W. Type A behavior pattern and coronary atherosclerosis. *Circulation,* 1978, *58,* 634–639.

Conley, M. J., Wechsler, A. S., Anderson, R. W., Oldham, H. N., Sabiston, D. C., and Rosati, R. A. The relationship of patient selection to prognosis following aortocoronary bypass. *Circulation,* 1977, *55,* 158–163.

Cook, W. W., and Medley, D. M. Proposed hostility and pharisaic-virtue scales for the MMPI. *Journal of Applied Psychology,* 1954, *38,* 414–418.

Dimsdale, J. E., Hackett, T. P., Hutter, A. M., Block, P. C., and Catanzano, D. M. Type A personality and extent of coronary atherosclerosis. *American Journal of Cardiology,* 1978, *42,* 583–586.

Dimsdale, J. E., Hackett, T. P., Hutter, A. M., Block, P. C. and Catanzano, D. M. Type A behavior and angiographic findings. *Journal of Psychosomatic Research,* 1979, *23,* 273–276.

Frank, K. A., Heller, S. S., Kornfeld, D. S., Sporn, A. A., and Weiss, M. D. Type A behavior pattern and coronary atherosclerosis. *Journal of the American Medical Association,* 1978, *240,* 761–763.

Hammermeister, K. E., DeRouen, T. A., and Dodge, H. T. Variables predictive of survival in patients with coronary disease. *Circulation,* 1979, *59,* 421–430.

Harris, P. J., Phil, D., Harrell, F. E., Lee, K. L., Dehar, V. S., and Rosati, R. A. Survival in medically-treated coronary artery disease. *Circulation,* 1979, *60,* 1259–1269.

Kloster, F. E., Kremkau, E. L., Ritzmann, L. W., Rahimtoola, S. H., Rosch, J., and Kanarek, P. H. Coronary bypass for stable angina. *New England Journal of Medicine,* 1979, *300,* 149–157.

Kolata, G. B. Consensus on bypass surgery. *Science,* 1981, *211,* 42–43.

Krantz, D. S., Sanmario, M. I., Selvester, R. H., and Matthews, K. A. Psychological correlates of progression of atherosclerosis in men. *Psychosomatic Medicine,* 1979, *41,* 467–475.

Review Panel on Coronary-Prone Behavior and Coronary Heart Disease. Coronary prone behavior and coronary heart disease: A critical review. *Circulation,* 1981, *63,* 1199–1215.

Sternbach, R. A. *Pain patients: Traits and treatment.* New York: Academic Press, 1974.

Whalen, R. E., Wallace, A. G. (by invitation), McNeer, J. F., Rosati, R. A., and Lee, K. L. The natural history of coronary artery disease: An update on surgical and medical management. *Transactions of the American Clinical and Climatological Association,* 1977, *89,* 19–33.

Williams, R. B., Haney, T. L., Lee, L. H., Kong, Y. H., Blumenthal, J. A., and Whalen, R. E. Type A behavior, hostility, and coronary atherosclerosis. *Psychosomatic Medicine,* 1980, *42,* 539–549.

Williams, R. B., Haney, T. L., Lee, K. L., Harrell, F. E., McKinnis, R. A., Blumenthal, J. A., Rosati, R. A., King, Y., and Sabiston, D. C. Psychosocial factors and pain relief in patients with CHD. Presented at American Heart Association, 53rd Scientific Sessions, November, 1980.

Wiltse, L. L., and Rocchio, P. D. Preoperative psychological tests as predictors of success of chemonucleolysis in the treatment of low-back syndrome. *Journal of Bone and Joint Surgery,* 1975, *57a,* 478–483.

Zyzanski, S. J., Jenkins, C. D., Ryan, T. J., Flessas, A. and Everist, M. Psychological correlates of coronary angiographic findings. *Archives of Internal Medicine,* 1976, *136,* 1234–1237.

Chapter 8

Post Myocardial Infarction

L. Howard Hartley

1. Introduction

This paper examines several aspects of behavioral science related to cardiac rehabilitation and emphasizes subjects the author believes are on the forefront of the field. Of special interest are behavioral factors which may influence the recovery process after a cardiac event. Although these comments may apply to many clinical states, this discussion emphasizes myocardial infarction.

Although outcome values of patients who have had a myocardial infarction are highly influenced by age and severity of disease, certain generalizations can be made. More than half of patients die within the first few hours after a myocardial infarction and before reaching the hospital. Of patients who survive to reach the hospital, approximately 15% will die before discharge. After discharge from the hospital, approximately 80% return to their usual way of life including previous employment. However, 20% become disabled. This discussion will cover the behavioral impact on the natural history of myocardial infarction with respect to survival and disability.

2. Behavioral Aspects of Myocardial Infarctions

The behavioral aspects of myocardial infarction are extremely important in all phases of the disease. These various phases are as follows:

Prehospital phase
 denial of illness
Coronary care unit
 anxiety
 relief

L. Howard Hartley • Cardiovascular Division, Brigham Women's Hospital, 75 Francis St., Boston, Massachusetts 02115.

exposure to stress-provoking situations
depression
Hospital ward
anxiety
depression
Posthospital convalescence
depression
fear of disability
vulnerability to misinformation

The prehospital phase of myocardial infarction is frequently characterized by denial (Hackett and Cassem, 1975). Many patients wait 4 to 6 hours before seeking help after developing the pain of cardiac infarction. Misdiagnoses of indigestion and musculoskeletal disorders are common, and patients delay seeking medical help while waiting for antacids or aspirins to provide relief. Denial is not restricted to uneducated or lower socio-economic backgrounds, and even physicians have delayed seeking medical help for several hours. The American Heart Association suggests that 50% of myocardial infarction deaths could be prevented if patients would seek help earlier.

After admission to the coronary care units, patients experience a variety of responses (Hackett, Cassem, and Wishnie, 1968). Anxiety intermingles with feelings of relief that medical care is available. However, the coronary care unit is frequently busy, and procedures being carried out on other patients may be disturbing. Even with this atmosphere, patients are usually relieved to be in an intensive care situation. Depression may appear in patients in the coronary care unit, although it usually occurs later.

After transfer from the coronary care unit to the hospital ward, patients may become anxious because of departure from the intensive care surroundings (Klein, Kliner, and Zipes, 1968), and depression may appear. The depression usually becomes prominent several days after departure from the coronary care unit (Hackett and Cassem, 1975). The approaching discharge from the hospital may also provoke anxiety.

During the posthospital convalescence, depression is again likely to appear or to worsen as the patient worries about the return to ordinary ways of life. Fears of disability and a sense of helplessness after departure from the hospital result in feelings of anxiety. Excessive concerns of relatives and friends and inappropriate advice may add to that anxiety.

Depression has a very direct relationship to the prognosis of patients. In observations of early studies, patients who are unable to cope with their illness are more prone to complications and death (Hackett *et al.*, 1968). A poor adjustment to the illness leads to a poor prognosis (Garrity and Klein, 1975). Increased cortisol and catecholamine excretion (Klein *et al.*, 1968) after a myocardial infarction may make patients more prone to complications. Fewer

Table 1

Characteristics of 172 Males in the Framingham Study Who Had Myocardial Infarctions in Relation to Reemployment

	Working (81%)		Nonworking (19%)
Age (years)	51.1	p .01	59.7
Education (years)	11.2		8.6
Neurocirculatory asthenia	10%		17%
AHA Function Class I	28%		33%
AHA Function Class II	68%		54%
AHA Function Class III	4%		12%
Congestive heart failure	12%		29%
Angina	45%		58%

patients who are depressed return to work compared to those who are better adjusted (Stern, Pascak, and Ackerman, 1977).

Although depression is a serious problem which is associated with poor prognosis, interventions can help. The use of psychotherapy counteracted depression in one study (Gruen, 1975). Depression is counteracted by using a variety of coping mechanisms, including denial. Exercise is an antidepressant and leads to lessening of depression symptoms in many patients (Hellerstein, 1968).

The ability to return to work is heavily influenced by behavioral factors. Certainly many individuals who are declared permanently disabled do not have obvious medical or physiological derangements which are different from patients who return to work (Fisher, 1970).

An examination of the characteristics of 173 males in the Framingham Study (Table 1) in relationship to reemployment after myocardial infarction revealed that 81% returned to work and 19% remained nonworking (Fisher, 1970). The only factor which was different between the two groups was age, which was significantly older in the nonworking population. Factors that determine whether a patient will return to work do not usually have an obvious medical or physiological origin. The use of psychological rehabilitation after myocardial infarction leads to greater social independence but not to more likelihood for return to work (Naismith, Robinson, Shaw, and MacIntyre, 1979).

3. Compliance

An activity is only effective if patients are willing to comply. The following data in Table 2 demonstrate compliance to exercise rehabilitation programs, ues are surprisingly low considering the severity of the illness and the popular perception of many that the programs will be of benefit.

Table 2
Compliance to Exercise Rehabilitation Programs

1968 Hellerstein	36 months	75%	(7)
1973 Cavanaugh	24 months	81%	(10)
1973 Sanne	30 months	39%	(11)
1975 Hartley	12 months	70%	(12)
1978 Oldridge	12 months	71%	(10)

The reasons why some patients comply and others do not are not completely understood. Differences between noncompliers and compliers in an exercise rehabilitation program were relatively few: Noncompliers had type A behavior, were inactive, and were smokers (Oldridge, Wicks, Hanley, Sutton, and Jones, 1978). Also, patients who have severe depression comply less frequently (Oldridge *et al.,* 1978). Many reasons are given for dropping out from exercise programs, but only 15% of patients dropped out because of medical reasons (Sanne, 1973). Reasons given by patients for noncompliance include practical difficulties and other unspecified problems which amount to more than half of the total number of cases.

Patient attitudes toward training programs are usually quite positive (Sanne, (1973). Patients believe that the exercise is not difficult to perform, that it is valuable for their health, and that activity is generally beneficial. Individuals who stop training perceive it as being less valuable and do not believe that exercise will improve their health. Perhaps it is this difference in perception that is pivotal. A number of factors can enhance compliance to behavioral therapy. Some of them are listed in Table 3:

Table 3
Factors Known to Affect Compliance to Behavioral Therapy

1. Instructions and information
2. Reinforcement
    social praise
    therapist approval
    involvement of spouse
    group involvement
3. Stimulus control
    teaching appropriate procedures
    self recording
4. Fading of therapist contact

Application of these factors to cardiac rehabilitation therapy has not been demonstrated to have positive effect; however, further research is (urgently) needed to assess these measures.

4. Physiological Correlates of Behavioral Events

Behavioral responses do correlate with a number of physiological measurements. Urinary catecholamines (Klein *et al.*, 1968), urinary sodium/potassium ratios (Gentry, Musante, and Haney, 1973), heart rate, blood pressure, and arrhythmias (Klein *et al.*, 1968) increase during myocardial infarction. Most of the dramatic responses occur in the coronary care unit or directly after leaving the unit, and fewer changes occur after the patient begins to gain confidence on the wards.

Psychological factors correlate with physiological factors during exercise conditioning. Following myocardial infarction the workload positively correlates with subjective feelings of fitness and inversely correlates with anxiety and subjective symptoms of heart disease (Prosser, Carson, Gelson, Tucker, and Neophylon, 1978). These behavioral–physiological correlates need further investigation.

5. Conclusions

a. Psychological factors have an important role to play in the recovery process after myocardial infarction.

b. Many psychological complications of myocardial infarction can be prevented or reversed by appropriate treatment.

c. Although certain clues have been discovered, the reasons for noncompliance of patients to therapies are not clear.

d. Physiological correlates of psychological evernts underscore the direct importance of these factors in patient outcome.

References

Fisher, S. Impact of physical disability on vocational activity: Work status after myocardial infarction. *Scandinavian Journal of Rehabilitation Medicine,* 1970, *65,* 2–3.

Garrity, T. F., and Klein, R. F. Emotional response in clinical severity as early determinants of six-month mortality after myocardial infarction. *Heart and Lung,* 1975, *4,* 730–737.

Gentry, W. D., Musante, G. J., and Haney, T. Anxiety and urinary sodium/potassium stress indicators on admission to a coronary-care unit. *Heart and Lung,* 1973, *2,* 875–877.

Gruen, W. Effects of brief psychotherapy during the hospitalization period on the recovery process of heart attacks. *Journal of Consulting and Clinical Psychology,* 1975, *43,* 223–232.

Hackett, T. P., and Cassem, N. H. Psychological management of myocardial infarction patient. *Journal of Human Stress,* 1975, *1,* 25–38.

Hackett, T. P., Cassem, N. H., and Wishnie, H. A. The coronary care unit: An appraisal of its psychological hazards. *New England Journal of Medicine,* 1968, *279,* 1365–1370.

Hartley, L. H. (personal observations)

Hellerstein, H. K. Exercise therapy in coronary disease. *Bulletin of New York Academy of Medicine,* 1968, *44,* 1028–1047.

Klein, R. F., Kliner, V. A., and Zipes, D. P. Transfer from a coronary care unit. *Archives of Internal Medicine,* 1968, *137,* 1680–1685.

Naismith, L. D., Robinson, J. F., Shaw, G. D., and MacIntyre, M. M. J. Psychological rehabilitation after myocardial infarction. *British Medical Journal,* 1979, *1,* 439–446.

Oldridge, N. B., Wicks, J. R., Hanley, C., Sutton, J. R., and Jones, N. L. Noncompliance in an exercise rehabilitation program for men who have suffered a myocardial infarction. *Canadian Medical Association Journal,* 1978, *118,* 361–364.

Prosser, G., Carson, T. P., Gelson, A., Tucker, H., and Neophyton, M. Assessing the psychological effects of an exercise training program for patients following myocardial infarction: A pilot study. *British Journal of Medical Psychology,* 1978, *51,* 95–102.

Sanne, Harold. Exercise tolerance and physical training of nonselected patients after a myocardial infarction. *Acta Medica Scandinavica Supplement,* 1973, *551.*

Stern, N. J., Pascale, L., and Ackerman, A. Life adjustment post myocardial infarction. *Archives of Internal Medicine,* 1977, *137,* 1680–1685.

Part IV

Prevention of Arteriosclerosis

Biobehavioral Factors in the Prevention of Arteriosclerotic Cardiovascular Disease

Redford B. Williams

The discovery in such large-scale prospective studies as the Framingham study that the presence of certain risk factors increased the subsequent risk of developing clinically apparent arteriosclerotic cardiovascular disease led to the hope and expectation that education of the public with regard to these findings would lead to changes in behavior and lifestyle that in turn would result in a reduced prevalence of the risk factors and, eventually, in the incidence of disease. While such public education efforts have undoubtedly played a role in the clear decrease in the incidence of strokes and the smaller, but still significant, decrease in the mortality rate associated with coronary heart disease over the past decade, it is nevertheless evident that much more remains to be accomplished. Since behaviors and lifestyles, whether they be cigarette smoking or failure to adhere to an antihypertensive regimen, are clearly involved in the continuing presence of the established risk factors, many leaders in biomedicine have been led to ask whether application of behavioral science knowledge and techniques might help to improve our efforts at prevention of arteriosclerotic cardiovascular disease through more effective approaches to risk factor reduction.

The research designed to answer these questions has only recently begun, and only the short- and long-term evaluation of these research efforts will enable us to reach final conclusions regarding the effectiveness of behavioral science approaches to risk factor reduction. It is already possible, however, to identify areas where sufficient work has been done to point the way to needed future investigation. Cessation of cigarette smoking is one area that has received extensive attention from behavioral scientists, and it appears on the basis of available data that behavioral science has much to offer in reducing

Redford B. Williams • Duke University Medical Center, Department of Psychiatry, Box 3416, Durham, North Carolina 27710.

the magnitude of this serious public health problem. In both California and in Finland primary prevention programs have been mounted employing both media and community organizations. While the analyses of the impact of these efforts is still incomplete, the preliminary findings are at least encouraging. Finally, the finding of decreased CHD rates among at least certain subsets of patients with treatment of mild hypertension in the Hypertension Detection and Follow-up Program has led to a renewed interest in the use of behavioral approaches to blood pressure reduction, both alone and in combination with pharmacologic treatment.

Despite the well-documented increased risk of both lung cancer and CHD among cigarette smokers and the massive campaigns to communicate these facts to the American public, more than 50 million Americans continue to smoke (Office of Cancer Communications, 1977). Even more disturbing is the observation that, although the rate of smoking has decreased among adults, among teenagers the rate of smoking has not declined appreciably—and, in fact, has even increased among female teenagers (U.S. Public Health Service, 1976). Moreover, the number of cigarettes smoked is increasing among those who do smoke (Fishbein, 1967), suggesting that the introduction of low tar and nicotine brands has resulted in increased consumption among addicted smokers to compensate for the loss of nicotine per cigarette.

Evaluation of the early efforts to reduce smoking behavior led to the conclusion that aversive techniques employing fear arousal (e.g., showing gory slides of emphysematous lungs) were generally ineffective in achieving smoking cessation (Bernstein and McAlister, 1976). It is likely that this failure results from the fact that while the long-term negative outcomes are clearly adverse, the short-term positive effects (e.g., stress reduction, good taste after a meal, avoidance of nicotine withdrawal symptoms) are likely to have more salience, especially for healthy younger smokers, leading to a greater "subjective expected utility" (Mausner, 1973) of continuing to smoke compared to going through the unpleasantness of kicking the habit.

The "rapid smoking" technique was one attempt to bring the aversive consequences of smoking more to the immediate attention of the smoker (Lichtenstein, Harris, Birchler, Wahl, and Schmahl, 1973). By having the client smoke rapidly to the point of nicotine intoxication, this technique apparently resulted in more favorable long-term abstinence rates than had earlier efforts. However, careful review (Evans, 1976) has revealed several methodological problems with even the best efforts at smoking cessation. First, the early high success rates with regard to follow-up at one year following the "rapid smoking" techniques have not been confirmed in subsequent studies, such that the "true" long-term abstinence rates appear to be no better than 25–30%. Second, all the subjects in the controlled outcome studies were volunteers and, hence, not necessarily representative of the addicted smoking population. Finally, the practice of excluding from analysis those subjects who did not complete the program may have artificially inflated the reported success rates.

Based upon the disappointing results of the reviewed studies aimed at cessation of smoking among addicted smokers noted above, the focus of anti-smoking research has recently shifted toward " . . . influencing pre-addictive smokers to curtail the incidence of smoking before they become addicted or nicotine-dependent, or to focus on preventing individuals from beginning to smoke in the first place" (Evans *et al.,* in press). Research into the origins of smoking behavior in children (Evans, 1976) indicated the importance of social influences to smoke from the media, peers, family, and other role models. Based on this observation, preliminary efforts have been undertaken using a "social inoculation" approach whereby preteens and teenagers are taught coping skills which hopefully will enable them to resist the many environmental influences upon them to smoke (Evans, 1976). A pilot application of this approach among seventh graders (Evans, 1976) was found to reduce smoking onset in the target groups by 50% compared to controls.

Several conclusions of the foregoing review of behavioral science approaches to smoking cessation are worth noting. First, application of such basic behavioral principals as those concerning schedules of reinforcement (i.e., long-term aversive outcome is unlikely to be effective in altering a behavior that is being maintained by immediate positive reinforcement) led to the conclusion that techniques based on fear arousal are unlikely to be effective—a conclusion supported by systematic evalution of such techniques. Second, a detailed behavioral analysis of the social determinants of onset of smoking behavior led to the identification of certain kinds of environmental influences to start smoking, which were then made the targets of a different approach based on social psychological principles. Thus, the behavioral science approach to smoking cessation has been a systematic one involving observation followed by hypothesis formation and testing and, eventually, evaluation of the outcome. While only long-term follow-up studies and replication in other populations will determine the effectiveness of the current social inoculation approach, it clearly represents a scientifically sound and rigorous effort to deal with what must be acknowledged as an extremely complex and resistant public health problem. Clearly, the current high levels of activity in this area are fully warranted.

The oldest and most comprehensive community-wide primary prevention program for arteriosclerotic cardiovascular disease was that instituted in the county of North Karelia, Finland, in 1972. A comprehensive review of the preliminary findings of this large-scale prospective study was presented to the Working Group on Arteriosclerosis in La Jolla on November 16, 1979 by the Project Director, Dr. Pekka Puska. That presentation provides the basis of this brief review. Of first importance in the initiation of this project was the recognition that Finland in general, and North Karelia in particular, experience the highest rates of CHD in the world. Moreover, the traditional risk factors of high blood pressure, cigarette smoking, and serum cholesterol elevations are highly prevalent in North Karelia. In response to expressions of community alarm at this CHD "epidemic," expressions which took the form of formal petitions to the central government, the North Karelia Project was begun in 1972. The program's activities were integrated with the service structure and social organization of the area, and a broad range of community actions was undertaken, consisting of use of the mass media and other information measures, provision of practical services (e.g., hypertension screening and follow-up), training of personnel, and attempts to modify certain aspects of the environment. Baseline and follow-up assessments of both levels of the targeted risk factors and morbidity and mortality were undertaken both in North Karelia and in another matched reference county to evaluate the impact of the interventions which were instituted.

Perhaps because the initial impetus for the project came from the local population, the participation levels and support enjoyed by the program remained at a high level throughout the project. There was a comparable reduction of from 52% to 33% in the rate of smoking among males in both North Karelia and the reference county. However, the number of cigarettes smoked showed a 10% greater decrease in North Karelia. Males in North Karelia showed a 4% greater reduction in serum cholesterol compared to the reference county. With regard to reduction in the prevalence of men with blood pressures in the hypertensive range, North Karelia showed a 40–50% greater decrease than did the reference county. To evaluate the net reduction in all targeted risk factors in North Karelia compared to the reference county, multiple logistic risk functions were computed for both communities, and North Karelia showed a greater reduction in this index of multiple risk factors—17% greater for men and 12% greater for women than the reference county.

With regard to morbidity and mortality outcomes, the rates of both first and recurrent myocardial infarctions had both decreased in North Karelia by the fifth year of the project period. The rates for recurrent MIs showed the greatest decline, perhaps indicating a larger effect for secondary prevention efforts. The incidence of stroke showed an even greater decline—30–40% by the end of the project period. While the cardiovascular disease disability pension rates were higher in North Karelia than in the reference county at the start of the project, by the fifth year there were on the average 15–20% fewer disability pensions in North Karelia—estimated to represent a $4 million saving over the five-year period. In terms of possible adverse effects of the intervention program in North Karelia, rather than any increase, there was observed a decrease in a wide variety of psychosomatic complaints, in days missed due to illness, and in overall emotional problems in North Karelia compared to the reference county.

These preliminary findings from the North Karelia Project are most encouraging, not only in terms of the feasibility of mounting large-scale, community-wide primary and secondary prevention programs, but also in terms of the effectiveness of such programs in reducing risk factors and various indices of morbidity and mortality. Outgrowths of the original project have included a nationwide smoking cessation series on Finnish television and a pilot project evaluating the use of community-wide reduction in salt intake for the primary prevention of hypertension.

In the United States, the Stanford "three communities" project evaluated the affects on risk factors of a mass media program for risk factor reduction, run both alone and in combination with more personal attention. A reduction in blood pressure similar to that in North Karelia was observed, although in California it seemed more a result of weight loss and salt reduction than of increased numbers of individuals on medication. As with the North Karelia study, in the Stanford project greater reductions were found in the numbers of cigarettes smoked than in the number of people abstinent, except in the high-risk group receiving personal attention in addition to the media campaign.

It should be emphasized that evaluation of intervention efforts such as those in the North Karelia and Stanford projects must be based on studies in the field, and, as such, are based on quasiexperimental designs. Such an approach is appropriate, since it is in the field that these various interventions must have their effect if the long-range goal of reducing the risk of arteriosclerotic cardiovascular disease is to be achieved. Since biomedical scientists have not had the extensive experience with such designs as have social psychologists and others in education and communication research, it is essential in the future that increased input in this critical area is obtained from the relevant behavioral sciences.

The Hypertension Detection and Follow-up Program (HDFP), comparing 5000 patients with mild hypertension treated in special hypertension clinics with 5000 similar patients referred back for regular care in their community, found treatment in the special clinics with stepped care achieved treatment goals more often than those under community care. For older patients there was a significant reduction in cardiovascular mortality in the stepped-care group (*Journal of American Medical Association,* 1979). Thus, it now appears that the large group of individuals with mild hypertension in the range of 90–104 mmHg diastolic would benefit in terms of reduced cardiovascular disease from more vigorous efforts to lower their blood pressure. To achieve such reductions may not be a simple matter, however—for the same reasons that it is not a simple matter to get people to stop smoking cigarettes. Compliance with antihypertensive regimens involves motivating the patient to endure in many cases immediately aversive consequences (e.g., side effects of the drugs, costs, and identification as being "sick") of treatment in order to achieve a reward (i.e., remaining healthy) that will not be realized in many cases until decades have passed. This area of compliance with medical regimens has received increasing attention of late (Haynes, Taylor, and Sackett, 1979) and will not be reviewed in detail here. Suffice it to say that there is clearly a great need for application of behavioral science knowledge and techniques in this area similar to that described above with regard to smoking cessation.

Besides the need for increased application of behavioral science approaches to the problem of compliance with antihypertensive regimens, the HDFP results also bring into sharper focus the need to investigate the application of behavioral approaches to the treatment of hypertension—particularly in the group of patients with mild hypertension, where behavioral approaches might be more effective than in the treatment of patients with more severe and longstanding blood pressure elevations, who often have end-organ changes. Recent review (Seer, 1979) of the extensive literature concerning the use of behavioral techniques in the direct treatment of high blood pressure suggests that while biofeedback and/or various relaxation strategies have not been shown to be effective in achieving blood pressure reductions sufficiently large to be clinically significant, a combination of these techniques is sufficiently promising to warrant further evaluation with provisions for a more comprehensive approach to assessment and training. The group of patients with mild hypertension would be a particularly appropriate target for such efforts. Besides direct efforts at lowering blood pressure, behavioral approaches could be applied in other ways to lower blood pressure by indirect means. For example, regular exercise and reduction of salt intake have both been found to result in blood pressure reductions. A comprehensive behavioral program to lower blood pressure can be envisaged which involves the simultaneous application of several behavioral techniques, including dietary modification for weight loss and salt reduction, exercise, and biofeedback/relaxation training.

All three of these elements of a combined behavioral approach could have immediately positive reinforcing properties, including increased physical fitness, improved appearance, and improved ability to cope with stress (through application of relaxation skills). Thus, the problems of noncompliance which deter so many patients from adhering to drug regimens might be less of a problem. Even if some patients eventually go on to require drug therapy for adequate blood pressure control, it is likely that participation in the behavioral program could help to improve compliance by: 1) reducing the dose of antihypertensive medication required for adequate blood pressure control; 2) providing the patient with skills training in assuming responsibility for health enhancing behavior; and 3) setting up a structure for regular monitoring by the patient of his own health status and the means for improving it.

In addition to improving our ability to treat mild hypertension, a systematic integration of biological and behavioral approaches to the regulation of blood pressure could also help to shed light on underlying mechanisms responsible for the initiation and maintenance of elevated blood pressure in certain subsets of patients (Schwartz, Shapiro, Redmond, Ferguson, Ragland, and Weiss, 1979). Clearly, there is potentially much to be gained by bringing behavioral-science knowledge and techniques to bear on the many problems associated with essential hypertension.

The foregoing review makes several recommendations for future research efforts directed toward applying behavioral-science knowledge and techniques to problems of prevention of arteriosclerotic cardiovascular disease:

1.   In view of the clear status of cigarette smoking as a risk factor, present efforts to evaluate behavioral approaches to smoking cessation and prevention of onset of smoking should continue with support at the highest possible levels.

2.   In view of the encouraging preliminary results of the North Karelia and Three Communities projects, continued community-based programs to reduce risk factors appear warranted. Such programs would benefit from increased behavioral-science input—especially with regard to the use of quasiexperimental designs, basic principles of behavior modification, and the role of psychosocial factors (e.g., life change stress, social supports, and Type A behavior pattern) in determining health outcomes. This latter issue deserves particular emphasis, in that the differential outcomes observed in association with various psychosocial factors (e.g., increased mortality among nonmarried or Type A CHD patients) if not assessed and controlled for could prevent our being able to adequately evaluate primary and secondary intervention programs.

3.   In view of the apparent importance of reducing blood pressure in patients with mild hypertension, based on the findings of the HDFP, intensified efforts should be undertaken to evaluate the effectiveness of behavioral approaches to blood pressure control in this large patient group. More specifically, behavioral-science knowledge and techniques should be brought to bear on problems relating to: a) compliance with antihypertensive regimens; b) indirect reduction of blood pressure through exercise, weight loss, and salt reduction; c) direct reduction of blood pressure through biofeedback and relaxation/meditation techniques; and d) the interaction between behavioral approaches and pharmacologic approaches to blood pressure reduction.

References

Bernstein, D. A., and McAlister, A. The modification of smoking behavior: Progress and problems. *Addictive Behavior,* 1976, *1,* 89–102.

Evans, R. I. Smoking in children: developing a social–psychological strategy of deterrence. *Journal of Preventive Medicine,* 1976, *5,* 122–127.

Evans, R. I., Henderson, A. H., Hill, P. C., and Raines, B. E. Current psychological, social, and educational programs in control and prevention of smoking: A critical methodological review. In A. M. Gotto and R. Panoletti (Eds.), *Atherosclerosis reviews,* Vol. 6, New York: Raven Press, 1979.

Fishbein, M. Consumer beliefs and behavior with respect to cigarette smoking: A critical analysis of the public literature. Report prepared for the staff of the Federal Trade Commission, 1967.

Haynes, R. B., Taylor, D. W., and Sackett, D. L. *Compliance in health care.* Baltimore: Johns Hopkins, 1979.

Hypertension detection and follow-up program of cooperative group: Five-year findings of the hypertension detection and follow-up program. *Journal of American Medical Association,* 1979, *242,* 2562.

Lichtenstein, E., Harris, D. E., Birchler, G. R., Wahl, J. M., and Schmahl, D. P. Comparison of rapid smoking, warm smoky air, and attention placebo in the modification of smoking behavior. *Journal of Consulting and Clinical Psychology,* 1973, *40,* 92–98.

Mausner, B. An ecological view of cigarette smoking. *Journal of Abnormal Psychology,* 1973, *81,* 115–126.

Office of Cancer Communications. *The smoking digest: Progress report on a nation kicking the habit.* Bethesda, Maryland: National Cancer Institute, 1977.

Schwartz, G. E., Shapiro, A. P., Redmond, D. P., Ferguson, D. C. E., Ragland, D. R., and Weiss, S. M. Behavioral medicine approaches to hypertension: An integrative analysis of theory and research. *Journal of Behavioral Medicine,* 1979, *2,* 311–363.

Seer, P. Psychological control of essential hypertension: review of the literature and methodological critique. *Psychological Bulletin,* 1979, *86,* 1015–1043.

U.S. Public Health Service. *Teenage smoking. National patterns of cigarette smoking, ages 12 through 18, in 1972 and 1974.* DHEW publication No. (NIH) 76-931. U.S. Department of Health, Education & Welfare, Public Health Service, National Institute of Health, 1976.

Chapter 9

Social Developmental Approaches to Deterrence of Risk Factors in Cardiovascular Disease

Richard I. Evans

The first thing behavioral science researchers entering the health field may discover is what many health professionals and biomedical scientists have long known: prevention of disease involves some form of life-style intervention. If psychologists have problems developing successful strategies for modifying life-styles of the mentally ill, delinquent youth, criminals, or those who are in the throes of poverty, certainly no fewer difficulties are encountered in developing strategies to effect the kinds of life-style modifications that are involved in preventing cardiovascular disease.

In order to illustrate a developmental approach to cardiovascular disease risk factor control, this report describes research conducted over the past few years by the University of Houston Department of Psychology, Social Psychology/Behavioral Medicine Research Group in collaboration with the Baylor College of Medicine National Heart Center—the theoretical and methodological underpinnings of it, and some of the models used in the investigation.

In the 1960s, research uncovered some interesting aspects of self-destructive or irrational behavior such as may be involved in behaviors related to cardiovascular disease. Our early research efforts, in collaboration with dental scientists, explored the possibility of overcoming unintentional self-destructive behavior by employing persuasive communications to stimulate more constructive behavior. The aim was to encourage people to brush and floss their teeth in order to prevent most tooth decay and gum disease. It became clear that arousing fear about the dangers of certain destructive behaviors rarely discouraged persons from starting or continuing these behaviors. Rather, general

Richard I. Evans • Department of Psychology, University of Houston, Central Campus, Cullen Boulevard, Houston, Texas 77004.

education programs appeared to be of only limited value in halting self-destructive behaviors. Very specific recommendations about health promotion or enhancing behaviors appeared to be much more effective (Leventhal, Singer, and Jones, 1965; Evans, Henderson, Hill, and Raines, 1979a; Evans, Rozelle, Noblitt, and Williams, 1975).

The 1964 Surgeon General's Report on Smoking and Health presented a challenge to behavioral scientists to develop strategies for modifying lifestyles. Changes in behavior appeared to be crucial to efforts to decrease morbidity and mortality related to smoking. But earlier studies, such as our recent reviews of the smoking literature (Evans *et al.*, 1979a; Evans *et al.*, 1979b) revealed the difficulties of altering deleterious behaviors. The circumstances of cigarette smoking, which involve addiction or dependence and have been integrated into our lifestyles in a complex manner, suggested the need to develop strategies that would influence children to resist the pressures to begin a high-risk behavior. Thus the focus of our research became deterrence or prevention of smoking rather than cessation of smoking in addictive or dependent smokers.

A survey of school programs dealing with prevention of smoking revealed (a) that these efforts were generally based on fear arousal; (b) the programs largely emphasized the future consequences of smoking, such as heart disease or cancer, and failed to recognize that teenagers tend to care more about the present than the future; (c) included films that failed to utilize previous research on effective use of media; and (d) evoked a response that was counterproductive. In light of these findings, we decided to undertake a long-term study in the Houston Independent School District (Evans, Rozelle, Maxwell, Raines, Dill, Guthrie, Henderson, and Hill, 1981). Exploratory work showed that peer pressure seemed to be a very important factor in the onset of smoking in teenagers. Interviews conducted before both a pilot study (Evans, 1976) and the long-term study (Evans, Rozelle, Mittelmark, Hansen, Bane, and Havis, 1978) with a large population of seventh graders suggested that peer pressure, models of smoking parents, and messages of the mass media (such as cigarette advertising) may, individually or collectively, outweigh the belief of children that smoking is dangerous. Although by the time they had reached seventh grade all the children believed smoking was dangerous (and some between ages four and eleven had spent time trying to persuade their parents to give up smoking), as they grew older, pressures of other sorts became superimposed on the fear of this behavior, and the fear became insufficient to prevent smoking.

A conceptual base for the Houston study was the social learning paradigm of Bandura (1977). Part of the interventions utilized in this investigation was directed toward training adolescents to become aware of individuals with whom they were in contact who model smoking behavior; this included parents, older peers, and personalities in mass media. Two findings in our research support the notion that modeling is an important factor in adolescent smoking. First, in junior high schools that were combined with senior high schools, the incidence of smoking was significantly higher than in junior high schools that were not directly joined with senior high schools; and second, if one's parents smoked, there was much greater likelihood that the child would smoke than if neither parent smoked; if an older sibling and parent smoked, there was a four times greater chance that the child would smoke (Evans *et al.*, 1979b).

Another conceptual base for the study was McGuire's (1969, 1974) cognitive communication model, which emphasizes verbal inoculation against the influence of persuasive communications. It was hypothesized that this model might be the basis for a promising intervention strategy at the behavioral level if the inoculation involved equipping students with techniques for saying "no" or resisting pressures to begin smoking. Our study was also predicated on the theoretical notion that if students can be nursed through the period during which they are particularly vulnerable to social influences to smoke (the junior high school years), they will be less likely to begin smoking. There is very little smoking begun in college or after age eighteen.

Using the strategy of inoculation against pressures to smoke, the project involved exposing students (seventh through ninth grade) to film and poster antismoking messages. Various film messages were presented to subjects at different times during the school year. Rather than relying upon adult authority figures as communicators, the films featured adolescents of approximately the same age as the target population, who presented the smoking information and role-played certain social situations in which the pressure to smoke is encountered. The roles of the students in the films were presented honestly, i.e., the student spokespersons stated that they had been asked to present the messages and to play certain roles.

Films presented on the first day included information about the dangers of smoking to health, and, most prominently, a section describing and illustrating peer pressure and its effect on smoking behavior. Other films recapped the first film and presented information about parental influence on smoking behavior, including a depiction of parental pressure to smoke and not to smoke, and children's modeling of parents' smoking behavior. Still other films recapped the first two films and presented information dealing with mass media pressures to smoke. These films included a pictorial analysis of such advertising techniques as artistically hiding the Surgeon General's warning on cigarette packages and appeals based on implied sexual attractiveness and popularity. The final films were a general recap of earlier films. The various treatment films of the pilot were either presented in their entirety (all four films presented), or absent (no treatment presented) in various experimental and control groups.

Following the films, students were asked for written and oral responses to questions. The experimenter distributed brief questionnaires for subjects' written responses. Four sets of questions were prepared and presented in conjunction with each film. The questions, which incorporated a quasi-roleplaying device of allowing the respondent to make decisions concerning whether or not to respond to social pressures to smoke, were formulated in such a way as to attribute motivation to resist pressures to smoke to persons who had seen the film, and to attribute ability to decide whether or not to smoke to persons subjected to smoking pressure.

Following each film presentation and accompanying a discussion designed to reinforce the messages in the film, a poster representing a scene from the film was displayed in the classroom. The posters served as a continuous reminder of the film messages.

The five dependent variables measured included smoking information, smoking attitudes, intention to smoke, and reported smoking behavior, as well as nicotine-in-saliva analysis (Horning, Horning, Carroll, Stillwell, and Dzidic, 1973), which was used as an objective measure of the presence or absence of smoking. The Horning test determines the amount of nicotine present in saliva samples by a mass spectrometric analysis. The results allow for inferences to be made about the degree of smoking behavior practiced by the subjects. Cost of operation of the mass spectrometer precluded analysis for each subject on each occasion, but saliva samples were collected from each subject on each occasion, and a sampling of specimens from each group was analyzed. This technique was used to increase the validity of self-reports of smoking. An earlier study (Evans, Hansen, and Mittelmark, 1977; Evans *et al.,* 1979b) found that when subjects learned from a short film that their saliva could be analyzed, their self-reports on smoking became more accurate.

In a ten-week pilot study, rates of onset of smoking in the treatment schools were significantly lower than the onset rates in the prepost test control schools. More than 18 percent of the control groups had begun smoking, while less than 10 percent in the experimental groups had begun smoking. (The small number of subjects already smoking in the various experimental groups precluded a statistical comparison of onset rates among the experimental groups and the control group.) In a follow-up three-year study (Evans, *et al.,* 1981) involving thirteen junior high schools, results indicated that those who gained information from films were smoking less than those who did not gain information. Using the criterion of smoking two cigarettes a day or more, seven percent fewer in the treatment schools were smoking this frequently than those in the control schools. These results suggest that such interventions may prove more useful in deterring smoking among junior high school students than merely instructing them by traditional methods about smoking. Perhaps most important, these findings suggest that various kinds of interventions may be effective, particularly if they have a reasonable conceptual base supported by data on the target audience about its perceptions of the determinants of smoking (Evans *et al.*, 1979a; Evans *et al.*, 1981). Other investigations (e.g., McAllister, Perry, and Maccoby, 1979; Hurd, Johnson, Pechacek, Bast, Jacobs, and Luepker, 1980) using variations of our developmental approaches have also reported promising results.

If behavioral science researchers have learned from their past experiences of working with applied problems, they will not enter the field of behavioral medicine and address such problems as developmental approaches to risk factor restriction with overly optimistic expectations. It is now well known that many of the optimistic promises of success made by some behavioral scientists as they entered the "War on Poverty" backfired, and individuals in the political system consequently became suspicious of their claims. Such attitudes may be reflected to this day in the lowered funding priorities for behavioral research. If individuals in the health care system are led to expect too much from the contributions of the behavioral science researcher, the results for behavioral medicine will be unfortunate.

Furthermore, on becoming involved in this new field, behavioral scientists face many critical entry problems. The field of medicine has been heavily exposed to the techniques of behavior modification, so much so that virtually all psychologists are now referred to as "behaviorists" by many physicians. Explaining the particular capabilities, interests, and values of an individual in any given field of psychology such as social psychology may often prove to be difficult. For example, the importance of the need for prefield investigations involving fairly basic psychological research needed to develop intervention techniques may not be easily communicated to many members of the biomedical community, who may overestimate the value of the techniques already in hand by the behavioral researcher.

The very rigor of evaluation procedures demanded by behavioral science researchers can be threatening to some of the health professionals who employ interventions at an essentially clinical level. In addition, as we move into an applied area, behavioral science researchers need to learn to recognize new types of methodological problems which arise when real world settings become the context of research. Although those of us in social psychology who are involved in field interventions characteristically conceive them with a critical eye toward rigorous evaluation, such projects are subject to occasional whims of fate, or the unavoidable impact of real world events on "controlled" research.

There may, however, be promise in attacking the problems of risk factors from a developmental point of view as illustrated by the deterrence of smoking research. The community of cardiovascular scientists should be encouraged to approach the whole area of lifestyle intervention as related to cardiovascular disease through collaborative efforts with behavioral scientists.

References

Bandura, A. *Social learning theory,* Englewood Cliffs, N.J.: Prentice-Hall, Inc., 1977.

Evans, R. I. Smoking in children: developing a social psychological strategy of deterrence. *Preventive Medicine,* 1976, *5,* 122–127.

Evans, R. I. Smoking in children and adolescents: Psychosocial determinants and prevention strategies. In *Smoking and health: A report of the Surgeon General.* Washington, D.C.: U.S. Government Printing Office (DHEW Publication No. [PHS] 79-50066), 1979b.

Evans, R. I., Henderson, A. H., Hill, P. C., and Raines, B. E. Current psychological, social, and educational programs in control and prevention of smoking: A critical methodological review. In A. M. Gotto and R. Paeoletti (Eds.), *Atherosclerosis reviews.* New York: Raven Press, 1979a.

Evans, R. I., Rozelle, R. M., Noblitt, R., and Williams, D. L. Explicit and implicit persuasive communication over time to initiate and maintain behavior change: A new perspective utilizing a real-life dental hygiene program. *Journal of Applied Social Psychology,* 1975, *5,* 150–156.

Evans, R. I., Hansen, W. B., and Mittelmark, M. B. Increasing validity of self-report of behavior in a smoking-in-children investigation. *Journal of Applied Psychology,* 1977, *62*(4), 521–523.

Evans, R. I., Rozelle, R. M., Mittelmark, M. D., Hansen, W. B., Bane, A. L., and Havis, J. Deterring the onset of smoking in children: Knowledge of immediate physiological effects and coping with peer pressures, media pressure, and parent modeling. *Journal of Applied Social Psychology,* 1978, *8,* 126–135.

Evans, R. I., Rozelle, R. M., Maxwell, S. E., Raines, B. E., Dill, C. A., Guthrie, T. J., Henderson, A. H., and Hill, P. C. Social modeling films to deter smoking in adolescents: Results of a three-year field investigation. *Journal of Applied Psychology,* 1981, *66*(4), 399–414.

Horning, E. C., Horning, M. G., Carroll, D. I., Stillwell, R. N., and Dzidic, I. Nicotine in smokers, non-smokers, and room air. *Life Science,* 1973, *13,* 1331–1346.

Hurd, P. D., Johnson, C. A., Pechacek, T., Bast, L. P., Jacobs, D. R., and Luepker, R. B. Prevention of cigarette smoking in seventh grade students. *Journal of Behavioral Medicine,* 1980, *3,* 15–28.

Leventhal, H., Singer, R. P., and Jones, S. Effects of fear and specificity of recommendation upon attitudes and behavior. *Journal of Personality and Social Psychology,* 1965, *2,* 20–29.

McAllister, A. L., Perry, C., and Maccoby, N. Adolescent smoking: Onset and prevention. *Pediatrics,* 1979, *63,* 650–658.

McGuire, W. J. The nature of attitudes and attitude change. In G. Lindzey and E. Aronson (Eds.), *The handbook of social psychology,* Reading, Mass.: Addison-Wesley, 1969.

McGuire, W. J. Communication-persuasion models for drug education: Experimental findings. In M. Goodstadt (Ed.), *Research on methods and programs of drug education.* Toronto: Addiction Research Foundation, 1974.

Chapter 10

Community-Based Intervention on Cardiovascular Risk Factors: Experiences from the North Karelia Project

Pekka Puska

1. Introduction

Research that ultimately aims at control of cardiovascular disease (CVD) has proceeded at many levels: pathological studies, animal studies, biochemical and physiological studies, clinical studies, and epidemiological studies. A number of epidemiological studies during the last two decades have made important contributions to present efforts for prevention of CVD. Several prospective follow-up studies (e.g., the Framingham Study) have identified the major independent risk indicators. Results from some monofactorial intervention trials have indicated the possible causal role of these factors, and several large multifactorial trials are under way at present.

The community study forms the important link between this research and its large-scale application in the society. It aims at studying whether and how the existing knowledge can be used to control this modern epidemic.

This report concerns the background, principles, and actual results with one such project, the North Karelia project in Finland, which obviously has been the first major community study. At present, several other studies following the same principles have been launched in a few other countries.

Finland has for several decades been faced with extremely high rates of CVD and especially CHD. This situation has not improved in spite of the increase in standard of living or in the quantity of health services. This has been the obvious background for a long tradition of previous research in the field of CVD in the 50s and 60s.

The highest disease rates were found in the Eastern section of the coun-

---

Pekka Puska • National Public Health Institute, Mannerheimintie 166, 00280 Helsinki 28, Finland.

try—in North Karelia, a mainly rural area with 180,000 inhabitants. Practically ten years ago the representatives of that region signed a petition to the Finnish government for urgent action "to reduce the extremely high cardiovascular rates." It was clearly expressed by the people that mere continuation of the academic research was not enough; instead, something had to be done, and immediately.

After that petition, the North Karelia project was designed and formulated (Puska, 1973). The project was fortunate to receive support and assistance for its planning from the World Health Organization and many international experts.

From the people's petition it was clear that a practical program to control CVD had to be started with its target whole area and its goal the greatest possible benefit for the population.

In the planning phase it became quite apparent that due to the chronic nature of CVD the potential for greatest success lay in primary prevention. With the enormous number of myocardial infarctions (over 1,000 acute cases per year in the population of 180,000), even a small relative decrease would mean a substantial number prevented in absolute terms.

The project planners were faced with numerous questions concerning the etiology and risk factors of CHD. However, due to the pressing need for action, decisions had to be made based on the best available knowledge. Fortunately, previous studies and many recommendations had highlighted the importance of three major established risk factors: smoking, high serum cholesterol, and high blood pressure, all of which were extremely prevalent in the area. Several other possible risk factors such as physical inactivity, obesity, and Type A behavior were not apparently prevalent in the area.

The extremely high level of serum cholesterol in the area is shown in the results from the Seven Country Study comparing the distributions among the North Karelian and the Japanese men (Keys, 1980). The distributions hardly overlap. Dietary differences between the two areas can well explain the difference in the cholesterol levels.

The ultimate design and strategy of the North Karelia project were further guided by some important considerations concerning the nature and operation of the risk factors, as had been learned through research in this field. Previous research from patient and animal studies to the follow-up studies had suggested the existence of the several risk factors. A natural continuation was the intervention studies. In the beginning of the 70s results from a few single-factor intervention trials, such as the Finnish Mental Hospital Study, started to be available (Turpeinen, Karvonen, Pekkarinen, Miettinen, Elosuo, and Paavilainen, 1979).

Since research had pointed out the multifactorial background of the problem as well as the synergism of risk factors, planning for trials aimed at multiple-risk-factor intervention had begun. But since such clinical trials must deal with thousands of people over a period of several years, and since they aim at influencing risk factors that are closely related to general lifestyles and the social and physical environment in the community, would it not be better to change the community instead of the individuals?

This last argument was further strengthened by the observation that in a typical high-risk community, the bulk of the severe disease cases comes not from the limited group of people with clinically high-risk-factor levels, but from the large population segment with slight elevation of several of the risk factors. Thus, the common approach of identification and intervention of high-risk people has severe limitations as a community strategy both from epidemiological as well as from behavioral and social points of view. These and other considerations led rather naturally to adoption of the community approach as the strategy in the North Karelia project.

2. The Program and its Evaluation

The major objective of the North Karelia project was reduction of mortality and morbidity, especially of major CVD, among the whole population but with special reference to middle-aged men. The central intermediate objectives were general reductions of the risk factor levels, including reduction of smoking, changes in dietary habits, and control of high blood pressure. At the same time, the program was viewed as a national pilot program to test this approach for possible nationwide use (Puska, 1973).

After the baseline surveys in 1972, a systematic intervention program was started. It was a comprehensive coordinated action where the activities were closely integrated with the existing service structure and social organization of the community. The aim was to have a popular community action by mobilizing the community itself to work for the set goals and to support this by a systematic and practical service structure.

The practical activities included especially the use of mass media, distribution of health education material, arrangements for practical services, training of local personnel, introduction of environmental changes, and the setting up of a practical information system for the management of the program.

The underlying intervention strategies took advantage of the initial health knowledge of the people and emphasized especially persuasion: (making changes for the sake of North Karelia and the community action) and teaching practical skills for change. Also, organization of social and environmental support for sustained changes was a central aim (McAlister, Puska, Salonen, Tuomilehto, and Koskela, 1981).

According to the plan, the original project was to evaluate this program for its first five-year period. The aim in the evaluation was assessment of 1) feasibility, 2) effects (risk factors, disease rates), 3) costs and 4) process and other changes associated with the program (Puska, Tuomilehto, Salonen, Nissinen, Virtamo, Björkqvist, Koskela, Neittaanmäki, Takalo, Kottke, Mäki, Sipilä, and Varvikko, 1981).

For the evaluation a matched reference area was chosen. At the outset in 1972, a baseline survey of a random population sample of some 11,000 people was carried out in the two areas (men and women, ages 25 to 59 years). In 1977 a similar survey was carried out for another cross-sectional sample of the same size. Strictly similar and standardized methods were used each time and in the two areas. The participation rates were over 90 on an average.

As the program effect was attributed, the observed change in North Karelia minus the respective change in the reference area was equal to the *net change*. In addition a number of other data sources were available or set up. These consisted especially of community-based disease registers and of existing mortality, hospital, and disability pension data.

3. Results

Table 1 gives the information on the risk factor levels and the actual disease rates in North Karelia, as measured at the outset of the program (in 1972).

Table 1
Baseline Situation in North Karelia of Risk Factor Levels (25–59 Years) and CVD Incidence (30–64 Years)

	Men		Women
Smokers (%)	52		12
Number of cigarettes (mean, smokers only)	18		10
Serum cholesterol (mean mg%)	269		265
Daily total fat consumption (mean, gr)	147		89
Daily total fat consumption (% energy)	38		35
PS-ratio		0.19	
Systolic blood pressure (mean, casual, mmhg)	147		149
Diastolic blood pressure (mean, casual, mmhg)	91		91
Prevalence of elevated blood pressure ($\geq$ 170 and/or 100%)	26		30
Annual incidence rate of acute myocardial infarction (per 1000)	13.8		2.6
Annual incidence rate of cerebrovascular stroke (per 1000)	2.9		2.5

Table 2

Reported Amount of Daily Smoking in North Karelia and the Reference Area in 1972 and 1977 According to Sex and Age Cohort

	Males	Females
1972		
North Karelia	9.9	1.3
Reference area	8.9	1.4
Difference	1.0**	−0.1
1977		
North Karelia	8.1	1.1
Reference area	8.1	1.3
Difference	0.0	−0.2
Net reduction in North Karelia		
Absolute	1.0*	0.1
Percent	9.8	8.0

*p < 0.05, **p < 0.01: statistical significance between North Karelia and the reference area (in 1972 and 1977) or between respective changes in the two areas (net reduction).

The feasibility of the program was good, as was the cooperation from the local services, and the general support by the community.

Comparison of the results from the baseline and five-year terminal surveys showed that in all aspects health behavior and risk factors changes took place in North Karelia in accordance with the objectives. Similar changes could be seen also in the reference area, but were, however, smaller than those in North Karelia. Tables 2, 3, 4, and 5 show the changes in smoking, selected dietary

Table 3

Net Reduction in North Karelia in Some Nutrition Behavior Variables from 1972 to 1977 (Percent)

Variable	Net Reduction in North Karelia (%)	
	Men	Women
Uses butter on bread	8.2	8.0
Uses butter in cooking	6.8	3.5
Uses at least 10 g of fat on a slice of bread	36.6	56.2
Uses high fat milk	10.9	5.1
Uses cream with coffee	8.1	18.4
Eats normally the visible fat of meat	7.0	13.3
Uses at least two lumps of sugar in a cup of coffee	4.3	−3.5

Table 4

Mean Serum Cholesterol Levels in North Karelia and the
Reference Area in 1972 and 1977 According to Sex and
Age Cohort (mg / 100 ml)

	Males	Females
1972		
North Karelia	269.3	265.3
Reference area	260.4	259.2
Difference	8.9***	6.1***
1977		
North Karelia	259.0	258.2
Reference area	261.2	255.1
Difference	−2.1	−3.1
Net Reduction in North Karelia		
Absolute	11.1***	3.0
Percent	4.1	1.2

***p < .001: statistical significance between the levels in North Karelia and those
in reference area (in 1972 or 1977) or between respective changes in the two
areas (net reduction).

Table 5

Mean Casual Systolic Blood Pressure Levels in North Karelia
and the Reference Area in 1972 and 1977 According to Sex
and Age Cohort (mmhg)

	Males	Females
1972		
North Karelia	147.3	149.4
Reference area	145.0	144.1
Difference	2.4***	5.3***
1977		
North Karelia	143.9	143.5
Reference area	146.8	145.4
Difference	−2.9***	−1.9**
Net Reduction in North Karelia		
Absolute	5.3***	7.2***
Percent	3.6	4.8

**p < 0.01, ***p < 0.001: statistical significance between the levels in North
Karelia and those in the reference area (in 1972 or 1977) or between the respective changes in the two areas (net reduction).

Table 6

Estimated Mean CHD Risk Scores in North Karelia and the
Reference Area in 1972 and 1977 According to Sex and
Age Cohort (Percent)

	Males	Females
1972		
North Karelia	4.1	3.3
Reference area	3.7	3.0
Difference	0.5***	0.4***
1977		
North Karelia	3.4	2.9
Reference area	3.7	2.9
Difference	−0.3***	0.0
Net Reduction in North Karelia		
Absolute	0.7***	0.4***
Percent	17.4	11.5

***$p < 0.001$: statistical significance between North Karelia and the reference
area (in 1972 or 1977) or between respective changes in the two areas (net
reduction).

habits, serum cholesterol, and casual systolic blood pressure results in North
Karelia, and the reference area according to sex.

Thus, for the male population, the following net reductions (effects) were
observed for the individual risk factors:

- smoking, 10%
- serum cholesterol, 11 mg/dl or 4%
- systolic blood pressure, 7 mmHg or 4%
- prevalence of high blood pressure, 44%

To assess the effects on the multiple risk factors simultaneously, a multiple
logistic function was applied and for each surveyed individual a CHD risk estimate was calculated based on his or her risk factor values. Thus, respective
means for the two areas and two points of time were obtained. The final outcome was a 17% net reduction among the men and a 12% net reduction among
the women in North Karelia in the general risk factor level during the program
period (Table 6).

Several further analyses of the changes have been completed. They essentially show that the changes and affects took place rather evenly among the
different sections of the population as regards age, socioeconomic status, or
initial risk.

The internal monitoring in North Karelia of morbidity and mortality
demonstrated clear reductions during the period. Among the male population

aged 30 to 64 years, the reduction in the incidence rates was 16% for acute myocardial infarctions and 38% for strokes, as measured by the community-based registers. The reduction in total mortality was practically exactly explained by the reduction in the CVD rates, with little change in other mortality statistics. When the mortality trends in the two areas were compared, some reduction was also seen in the reference area. Thus, within this period no statistically significant difference could be observed, in spite of clear reduction in North Karelia.

Disability payments are given in Finland by a National Pensions Institution after a doctor's certification. The number of CVD disability pensions given increased since 1968 in a parallel way in North Karelia and the reference area. After 1972, the increase in North Karelia slowed down compared with the reference county, and an approximate 15% net reduction in CVD disability occurred. The trends with non-CVD disability cases developed in a very similar way during the period in the two areas.

The questionnaires in the surveys contained a great number of questions on possible emotional problems concerning anxiety, stress, self-reported health and illness days, and other symptoms. In most items the development in North Karelia was positive, and in no instance less satisfactory than that in the reference area. When a comprehensive index of these reported emotional problems was formed, a net reduction of 6% among men and 10% among women was observed. The trend persisted even when the analysis was restricted for the high-risk groups alone.

The total direct costs of the project were approximately 1.7 million U.S. dollars. Estimation of the costs of the program for the community, a complicated task, indicated that the extra costs for the health and other community services were small, to a great extent due to the fact that the program mainly introduced a more systematic use of the existing resources, integrated simple preventive measures in its daily services, and made better use of nurses and paramedical personnel. Thus, the overall costs of the project were obviously limited and matched well with the obvious savings that occurred (a 4 million U.S. dollars savings for the government in reduced disability pension awards alone).

4. Conclusions

The results of the study are: 1) Good feasibility of the comprehensive community program; 2) Good cooperation with the local population and health and other local personnel; 3) Clear effect of the lifestyles and biological risk factors (changes in the program area significantly greater than in the reference area); 4) Incidence and mortality rates in the area reduced, but due to reduction in the reference area (probably in the whole country), no significant net effect in mortality could be demonstrated; 5) Net reduction in CVD and all disability pension rates in the program area; 6) Relatively low direct costs of the program to the community and obvious savings; and 7) Absolute and net reduction in reported emotional problems in the program area.

To these reported findings can be added a general satisfaction with the program as expressed by the population, local health personnel, and decision-makers.

On the one hand, it is felt by some that five years is obviously long enough to assess permanent lifestyle and risk factor changes. On the other hand, only continued follow-up will show the future trends and give a real picture on possible mortality effects.

Thus, both the follow-up and most of the other established activities continue in the area. At the same time, however, great national interest and numerous national applications have taken place. Preliminary information of a ten-year follow-up has indicated that the risk factor levels have further declined and that the net reduction has been maintained. Preliminary information is also available about further reduction in coronary mortality, indicating significantly more than in the rest of the country. It is obvious that the project has contributed to general adoption of the idea that the modern epidemic of CVD can be controlled just as many previous public health problems have been controlled. This time the problem seems to be even more complex and a longer time span must be provided for. Associated with the clearly positive development in North Karelia is the observation that during the last few years the national CHD mortality rule has also started to decrease in Finland.

References

Keys, A. *Seven Countries, a Multivariate Analysis of Death and Coronary Heart Disease.* Cambridge: Harvard University Press, 1980.

McAlister, A., Puska, P., Salonen, J. T., Tuomilehto, J., and Koskela, K. Theory and action for health promotion: illustrations from the North Karelia project. *American Journal of Public Health,* 1982, *72,* 43–50.

Puska, P. The North Karelia project—an attempt at community prevention of cardiovascular disease. *WHO Chronicles,* 1973, *27,* 55–58.

Puska, P., Tuomilehto, J., Salonen, J., Nissinen, A., Virtamo, J., Björkqvist, S., Koskela, K., Neittaanmäki, L., Takalo, T., Kottke, T., Mäki, J., Sipilä, J., Sipilä, P., and Varvikko, P. The North Karelia project: evaluation of a comprehensive community program for control of cardiovascular diseases in 1972–77 in North Karelia, Finland. *WHO Monograph,* in press.

Turpeinen, O., Karvonen, M. J., Pekkarinen, M., Miettinen, M., Elosuo, R., and Paavilainen, E. Dietary prevention of coronary heart disease: The Finnish mental hospital study. *International Journal of Epidemiology,* 1979, *8,* 99–118.

Chapter 11

Community Approaches to Risk Factor Reduction: The Stanford Project

John W. Farquhar

---

1. Theoretical Basis

A major theoretical basis for community risk reduction studies is contained in Bandura's social learning theory, which describes the reciprocal relationship between environment, behavior, and cognitive factors (Bandura, 1969; Bandura, 1977; Bandura, 1978). Translated into practical terms, this theory states that the artificial environments of clinics and doctors' offices, the province of most controlled clinical trials, do not have the potential for learning and maintenance of learning that is possible within the home, worksite, and other areas of community environments. Therefore, community studies can furnish a natural milieu for full testing of multifactor educational methods for achieving lifestyle change.

2. Utility of Community Studies

Community studies serve the following purposes:

1. To furnish an optimal environment for testing efficacy of educational methods of achieving lifestyle change.
2. To test hypotheses of postulated clinical benefit of change in risk factors. (It follows that if community education can enhance the effectiveness of learning, the primary-risk-factor hypothesis testing usually better done in clinic environments may need testing in communities.)
3. To test the generalizability of any results obtained in controlled clinical

---

John W. Farquhar • Stanford Heart Disease Prevention Program, Stanford University Medical School, Stanford, California 94305.

trials in selected volunteers. A variety of field trials may allow the "real world" utility of any results to be assessed more adequately.

4. To determine costs and benefits of generalizable risk reduction methods, including the evidence for interaction among treatment elements. These studies will furnish policy makers with needed data.

5. To furnish evidence for the process by which change occurs and the causal links between education and risk factor change. The multifactor nature of cardiovascular disease risk factors and the links between habits and their physiological consequences furnish a rich challenge for evaluation within community studies.

6. To allow study of the determinants of mass adoption of health innovations.

7. To allow study of the role of social networks in disease or precursors of disease and of the effect of changing social networks on the disease or its precursors.

8. To allow study of maintenance of change, including the role of social support.

9. To furnish a laboratory for needed studies in community organization for health.

10. To allow the study of methods for health professional education.

11. To allow study of youth education in healthy living in a setting where the family and social networks, in addition to the school setting, can be mobilized.

12. To allow study of worksite health programs in settings where the family and community support structures can be mobilized.

These manifold utilities have both practical and theoretical roles as befits studies done under the banner of field testing. In summary, community studies are a necessary complement to basic clinical research studies and are needed in order to achieve a natural empirical basis for policy decisions that can lead to control of cardiovascular diseases and of any other chronic disease that has important psychosocial roots.

3. The Stanford Three Community Study

3.1. History

The study's planning began in 1970 and its prefield operations in 1971 (Meyer and Henderson, 1974). Field operations began in 1972 and were completed in 1975 (Stern, Farquhar, Maccoby, and Russell, 1976; Farquhar, Wood, Breitrose, Haskell, Meyer, Maccoby, Alexander, Brown, McAlister, Nash, and Stern, 1977; Maccoby, Farquhar, Wood, and Alexander, 1977;

Meyer, Nash, McAlister, Maccoby, and Farquhar, 1980). The project began as an outgrowth of collaboration between members of Stanford's School of Medicine (Drs. J. W. Farquhar, P. D. Wood, M. P. Stern, W. Haskell, and B. W. Brown, Jr.) and members of Stanford's Institute for Communication Research (Drs. N. Maccoby, H. Breitrose, D. Roberts, and Ms. J. Alexander).

These individuals developed a scientific consensus that the medical, epidemiologic, and behavioral evidence pointed to community studies as a timely addition to cardiovascular disease control for the reasons outlined above.

3.2. Design

The study was based on the assumption that an educational program delivered largely through the relatively inexpensive mass media was deserving of a trial. Secondly, we assumed that supplemental face-to-face instruction might be needed to achieve lasting and important degrees of change considering the recalcitrant nature of the health habits underlying cardiovascular disease risk factors. This highly exportable and low-in-cost method could thus be compared with one that has a much better chance of being successful although perhaps too expensive and not readily exportable. The educational part of the study was carried out in two Northern California towns, Watsonville and Gilroy, with a third town, Tracy, serving as control. A large sample of adults aged 35–59 was surveyed yearly for risk factors as well as knowledge and attitude. The two towns receiving education were given a symmetrical exposure to printed booklets, radio, television, and newspaper material. Only in Watsonville did we test the "gold-standard" of supplemental face-to-face instruction given to a small subset of the population.

3.3. Results

The changes seen in risk factors were not only considerably greater than those expected in the media-plus-intensive-instruction condition, but were surprisingly well maintained in the media-only condition (Farquhar *et al.*, 1977; Meyer *et al.*, 1980; Williams, Fortmann, Farquhar, Mellen, and Varady, 1981). The media were successful in affecting changes in body weight, blood pressure, and cholesterol levels sufficient to achieve a 24% difference in estimated future risk of coronary heart disease (Farquhar *et al.*, 1977) (see Figure 1). However, very few changes relative to the control occurred in smoking rates in the media-only condition, indicating that extra attention was needed for this particular risk factor. In the intensive-media-plus-instruction group, 50% of the smokers were not smoking after three years compared to a 15% drop in the control group.

The lessons learned from the Three Community Study led us toward a more ambitious project, the Stanford Five City Project.

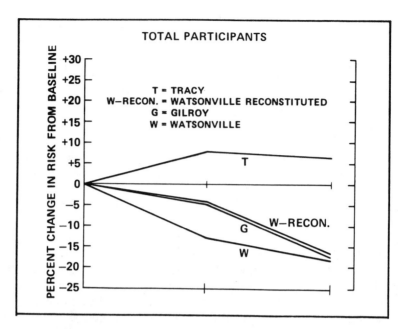

Figure 1. Percent change from baseline (0) in risk of coronary heart disease after one and two years of health education in various study groups from their communities. W and G received education, and T served as a control. W-RECON and G represent the media-only condition. W contains a subset of intensively instructed individuals in the sample. Reprinted with permission of *Lancet*.

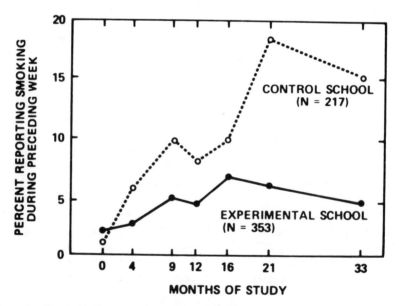

Figure 2. Changes in the reported prevalence of weekly smoking from longitudinal observation of two seventh grade study cohorts. Reprinted with permission of *Journal of Behavioral Medicine*.

4. The Stanford Five City Project

4.1. History

The core of investigators from the Three Community Study (Drs. J. W. Farquhar, N. Maccoby, P. D. Wood, B. W. Brown, Jr., W. Haskell, and Ms. J. Alexander) was joined by Drs. W. S. Agras and C. B. Taylor of Stanford's Department of Psychiatry, Dr. C. Thoresen of Stanford's School of Education, Drs. S. Hulley and S. P. Fortmann of Stanford's Department of Medicine, and Dr. E. W. Rogers of Stanford's Institute for Communication Research.

4.2. Design

These scientists planned a larger project to which community organization, youth education, and health professional education were added. In addition, the study was designed to include cardiovascular disease incidence, measured by community-wide monitoring of all new fatal and nonfatal episodes of coronary heart disease or stroke (Farquhar, 1978). The study was designed to run in the field for eight years and began the first year of six scheduled years of education in the Spring of 1980.

4.3. Results

No data from the study proper are as yet available. However, a promising pilot study on smoking prevention in 7th grade students has been completed (Telch, Killen, McAlister, Perry, and Maccoby, 1982) and Figure 2 displays the results of this study.

These exciting results indicate that peer group influence can successfully retard the onset of smoking in young people, and the results give us considerable hope that similar results will be obtained when such methods are applied within the context of a community-wide study. We anticipate, for example, a reciprocal and synergistic benefit between parents and children in health matters as we plan programs for schools and families.

5. Summary

Field demonstration studies are an essential component of cardiovascular disease control and are therefore of urgent priority. Fortunately, in addition to the Stanford Five City Project, two other large projects are now fully funded by the NIH and are banded together in a confederation of shared evaluation methods. These are the Minnesota Heart Health Study, under the direction of Dr. Henry Blackburn and colleagues, and the Pawtucket Heart Health Study, under the direction of Dr. Richard Carleton and colleagues. A fourth study, the Lycoming County Study, is jointly directed by Drs. Paul Stolley and Albert

Stunkard of the University of Pennsylvania and is supported by the State of Pennsylvania. The four studies form a composite of community-based, multifactor health education that will undoubtedly contribute to an understanding of the many important questions that remain unanswered in our search for cost-effective prevention and control of cardiovascular disease. Our relationships to sister projects in other countries, notably the North Karelia project in Finland, are also strong, and it is clear that international cooperation in community studies will assist medical science to achieve its goals more rapidly (McAlister, Puska, and Jyde, 1978; Puska, Tuomilehto, Salonen, Nissinen, Virtamo, Björkqvist, Koskela, Neittaanmäki, Takalo, Kottke, Mäki, Sipila, and Varvikko, 1981).

References

Bandura, A. *Principles of behavior modification.* New York: Holt, Rinehart and Winston, 1969.

Bandura, A. *Social learning theory.* Englewood Cliffs, N.J.: Prentice-Hall, 1977.

Bandura, A. The self system in reciprocal determinism. *The American Psychologist,* 1978, *33,* 344–358.

Farquhar, J. W. The community-based model of life-style intervention trials. *American Journal of Epidemiology,* 1978, *108,* 103–111.

Farquhar, J. W., Wood, P. D., Breitrose, H., Haskell, W. L., Meyer, A. J., Maccoby, N., Alexander, J. K., Brown, B. W., Jr., McAlister, A. L., Nash, J. D., and Stern, M. P. Community education for cardiovascular health, *The Lancet,* 1977, *1,* 1192–1195.

Maccoby, N., Farquhar, J. W., Wood, P. D., and Alexander, J. K. Reducing the risk of cardiovascular disease: Effects of a community-based campaign on knowledge and behavior. *Journal of Community Health,* Winter 1977, *3,* 100–114.

McAlister, A., Puska, P., and Jyde, J. Mass communication of cessation counselling: Combining television and self-help groups. *World Smoking and Health,* 1978, *3,* 28–32.

Meyer, A. J., and Henderson, J. B. Multiple risk factor reduction in the prevention of cardiovascular disease. *Preventive Medicine,* 1974, *3,* 225–236.

Meyer, A. J., Nash, J. D., McAlister, A. L., Maccoby, N., and Farquhar, J. W. Skills training in a cardiovascular health education campaign. *Journal of Consulting and Clinical Psychology,* 1980, *48,* 129–142.

Puska, P., Tuomilehto, J., Salonen, J., Nissinen, A., Virtamo, J., Björkqvist, S., Koskela, K., Neittaanmäki, L., Takalo, T., Kottke, T., Mäki, J., Sipila, P., and Varvikko, P. The North Karelia Project: Evaluation of a comprehensive community program for control of cardiovascular diseases in 1972–77 in North Karelia, Finland. *Public Health In Europe, WHO/ EURO Monograph Series,* Copenhagen, World Health Organization Regional Office for Europe, 1981.

Stern, M. P., Farquhar, J. W., Maccoby, N., and Russell, S. H. Results of a two-year health education campaign on dietary behavior: The Stanford Three Community Study. *Circulation,* 1976, *54,* 826–833.

Telch, M. J., Killen, J. D., McAlister, A. L., Perry, C. L., and Maccoby, N. Long-term followup of a pilot project on smoking prevention with adolescents. *Journal of Behavioral Medicine,* 1982, *5,* 1–8.

Williams, P. O., Fortmann, S. P., Farquhar, J. W., Mellen, S., and Varady, A. A comparison of statistical methods for evaluating risk factor changes in community-based studies: An example from the Stanford Three Community Study. *Journal of Chronic Diseases,* 1981, *34,* 565–571.

Part V

Resources, Manpower, and Training Programs

Chapter 12

Biobehavioral Resources, Manpower, and Training Programs

Joseph D. Matarazzo

1. Introduction

Probably the single most important development which helped the scientific community to recognize the role of behavioral factors in health and illness and the need to address this knowledge in future training and research was the scientific knowledge accumulated between 1960 and 1980 on the role of behavioral and environmental factors, including an individual's *lifestyle,* in his or her health. Thus, examples cited in the recently published *Report of the Working Group on Arteriosclerosis of the NHLBI* (1981)(1) include the following:

1. "One of the most complex problems in atherosclerosis is the association of risk and disease with human behavior, personality, and social and cultural influences. Principles of conditioning to social learning have been shown to be operative in behavior patterns which influence cardiovascular risk factors, adherence to medical regimens, and outcomes of treatment for cardiovascular disease. Although behavioral characteristics are clearly important in determining habits of diet, smoking, and exercise, they also exert independent effects through mechanisms *that are not as clear as are established biochemical and physiological pathways of atherogenesis*" (emphasis added here) (p.30).

2. "Basic neurobiologic studies have indicated mechanisms whereby behavioral processes influence metabolic and hemodynamic processes. Also, associations have been established indicating possible links between brain function, behavior, neuroendocrine and other physiological processes, and atherosclerotic cardiovascular disease. However, the mechanisms whereby psychosocial and behavioral processes influence atherosclerosis are still largely unknown.

Joseph D. Matarazzo • Oregon Health Sciences University, Portland, Oregon 97201.

"Research is needed on the physiological and biochemical links between behavioral processes and atherosclerosis. In particular, more research is needed on neurochemical, neuroendocrine, and metabolic functions which may influence atherosclerosis" (p. 34).

3. "Optimal treatment of cardiovascular diseases frequently involves combined medical, surgical, pharmacological, and psychosocial approaches. Clinical experience suggests that behavioral effects are involved in all of these approaches, designed to influence lifestyles, cardiovascular risk factors, adherence to medical regimens, and outcomes of treatment. Psychosocial and behavioral characteristics before and during active intervention need to be assessed with respect to modifying behavior for both primary and secondary prevention of the atherosclerotic diseases" (p. 32).

4. "Certain patterns of behavior seem to enhance the prospect for developing clinical manifestations of atherosclerosis. There is now extensive evidence indicating that individuals who respond to challenges with great competitive drive, hostility, and impatience (frequently called coronary-prone or Type A, behavior pattern) are more likely to develop coronary heart disease than individuals who do not respond in this way. Data from a few studies also indicate that major changes in occupation or residence, bereavement, or other stressful life events are associated with increased risk of clinical manifestations of coronary heart disease" (p. 18).

5. "The association between close interpersonal ties and a relative reduction in incidence and prevalence of atherosclerotic cardiovascular disease has been suggested in several epidemiologic studies involving several different behavioral measures in several different populations. However, the mechanisms of this possible protective effect are unknown.

"Research is needed to examine interrelationships among a range of behavioral factors—including social networks—and their influence on such socially determined factors as eating, drinking, and smoking habits; psychological factors such as levels of self-esteem and coping styles; and physiological factors such as circulating levels of catecholamines, secretion of cortisone, and other neuroendocrine substances" (p. 34).

6. "Research personnel needs: Behavioral scientists to study social, psychological, and behavioral mechanisms influencing the basic process of atherosclerosis, behavioral procedures for clinical treatment and rehabilitation, and social influences important in preventing atherosclerotic cardiovascular disease" (p. 35).

7. "Interdisciplinary research depends upon training young investigators to expand the scope of inquiry beyond their primary discipline. Because of the need for scientists with experience in biobehavioral research, programs have been set up for training scientists in the psychological and behavioral aspects of atherosclerosis. Professional development and training also has been

enhanced by interdisciplinary workshops, conferences, and colloquia sponsored by the National Heart, Lung, and Blood Institute.

"Training programs in biobehavioral research should be expanded to include preventive cardiology with strong biobehavioral components. In addition, training programs in cardiology should be strongly encouraged to develop a biobehavioral component. Also, the National Heart, Lung, and Blood Institute should continue to sponsor interdisciplinary workshops, conferences, and colloquia in an effort to enhance the utilization of behavioral science in research on atherosclerosis" (p. 36).

2. Research Training Program Goals, 1982 to 1987

As we move into the 1980's, advances in a number of scientific areas related to cardiovascular function, coupled with a number of demographic, economic, and social factors, have combined to change the pattern of training and employment for doctoral-level behavioral scientists. Graduate enrollment in the behavioral sciences give evidence of having levelled off and are projected to decline throughout the 1980s. Employment opportunities for new Ph.D.s, once centered primarily in academic and research institutions, are being outpaced by employment in new, less traditional settings (industry, government, community agencies, etc.).

However, as a seemingly paradoxical development, even in the face of these upheavals in what had been for decades the "traditional pattern" of training and employment, advances in research knowledge have occurred in selected areas such that *more* rather than less research training may be needed to meet the needs spawned by these new fields of knowledge. The area of interface between behavior and health has emerged to capture the attention of the scientific communities of the several disciplines involved. This interface is variously known as behavioral medicine, health psychology, behavioral health, behavioral cardiology, and preventive cardiology, among others. As an area requiring the attention of researchers, teachers of future researchers, and ultimately the skills of applied practitioners and educators, these newly emerged fields have interrelated with medical psychology, medical sociology, epidemiology, nutrition, medical anthropology, and the basic biomedical and applied clinical medical sciences with which each of these behavioral sciences interfaces.

In 1976 the *National Academy of Sciences' Committee on a Study of National Needs for Biomedical and Behavioral Research Personnel* and its *Panel on Behavioral Science* (Riecken and Wyngaarden, NAS Report, 1978) (2) recommended a major reorientation of federal support toward encouraging postdoctoral study while reducing predoctoral support under the auspices of the NRSA authority. These 1978 recommendations were made in response to the then changing employment conditions for young doctoral scientists and to the emerging areas of innovative research in the behavioral sciences in need of personnel development. This 1978 NAS Report identified (pp. 73–74) the area of *behavior and health* as an emerging area of national need and recommended that increased emphasis be placed on predoctoral and postdoctoral training in the behavioral sciences to increase the number of investigators in this area. This 1978 recommendation was put forth just as the NHLBI began a series of initiatives which, over the next three years, both helped to systematize the accumulated but disparate scientific knowledge reviewed above, and helped a number of research training centers evaluate their local capacities to offer predoctoral and postdoctoral training in interdisciplinary areas involving biomedical and behavioral components. Because, among other reasons, so much new knowledge had accumulated since its published 1978 report, a brief 2–3 years later (1980), the *NAS* sponsored a *Workshop on Establishing Research Training Programs in Behavior and Health* (Ebert-Flattau, National Academy Press, 1981) (3) at the *Institute of Medicine* to assess what had transpired in the three years intervening since the 1978 NAS Report.

Among other conclusions, this 1981 follow-up NAS Report concluded (pp. vii and viii) that:

(1) "A greater number of older, more mature individuals, in the largest proportion women, are seeking graduate training in the behavioral sciences at this time. Often from allied health disciplines, these persons have enhanced behavioral science programs by virtue of their interest and orientation."

(2) "The market for behavioral scientists remains particularly good for those with quantitative skills. There appears to be a growing demand for such persons in nonacademic settings."

(3) "The area of behavior and health is not presently a unified field. While a number of textbooks have been written on the subject, the use of such books and the content of graduate training vary tremendously from one institution to another. Health behavior research is clearly at a stage of dynamic growth."

(4) "There is a need for postdoctoral training in the area of behavior and health. Such training completes the research skills and subject knowledge acquisition begun at the predoctoral level."

(5) "There is also a need for predoctoral support in health behavior research training, especially in those institutions which have strong programs of research in place. As far as recommendations for appropriate predoctoral/postdoctoral support through the NRSA authority, conferees agreed that it is too early to judge what combination of awards is necessary for this area at this time."

(6) "There may be a special need to provide predoctoral institutional training grants to behavioral scientists teaching in health professional schools to permit a continuity in program development."

During the next decade this nation will need many more highly-trained behavioral science research personnel to exploit the new knowledge generated between 1960 and 1980, particularly in the following areas which hold great promise:

1. Research on the role of behavior in such lifestyle and risk factors as:
   a. smoking
   b. type A behavioral pattern
   c. hypertension
   d. obesity
   e. poor diet
   f. lack of exercise
   g. stress
2. Research on compliance and adherence
3. Research to develop animal models of the role of social stressors and conditioning in sudden death
4. More general research on how to prevent disease and promote cardiovascular health in currently healthy people

3. Current Research Personnel Demand

The need for biobehavioral scientists is well-documented. More than a decade ago, the NHLBI predicted that 120 new cardiovascular biobehavioral scientists per year would need to be trained by 1972. This goal was overly optimistic—even now *in 1982,* NHLBI is supporting only 58 predoctoral and postdoctoral trainees in institutional programs, a figure which does not come close to meeting goals appropriate a decade ago. These 58 trainees are distributed throughout the academic programs listed in Table 1.

The behavioral scientists in these programs receive training in such areas as cardiovascular physiology, pathophysiology, neurochemistry, and selected aspects of clinical cardiology. Likewise, the cardiologist in training is taught selected aspects of behavioral science such as relevant areas of personality theory; methods of individual behavior change through biofeedback and behavior modification; social learning theory as it pertains to community intervention studies; the role of health beliefs as determiners of health-related behavioral change; the role of life events and stress on physiological functions and techniques of stress management, to name but a few. To synthesize accumulated knowledge, for example, on the roles of lifestyles, neuropeptides, and endor-

Table 1

Doctoral and Postdoctoral Research Training Programs in Cardiovascular Behavioral Medicine Supported by the National Heart, Lung, and Blood Institute

Institution	Type	Training Director
Cornell Medical College	Two-year postdoctoral training in central nervous system control of circulation	Donald J. Reis, M.D.
Johns Hopkins University	Pre- and postdoctoral multidisciplinary training in disciplines related to health	David Levine, M.D., M.P.H., Sc.D.
Stanford University	Multidisciplinary postdoctoral training in cardiovascular disease prevention	John W. Farquhar, M.D.
University of California, Berkeley	Pre- and postdoctoral training in cardiovascular disease etiology and prevention	S. Leonard Syme, Ph.D.
University of Houston	Pre- and postdoctoral training in social psychology and cardiovascular disease	Richard I. Evans, Ph.D.
University of Minnesota	Predoctoral training in human behavior and cardiovascular disease	Russel V. Luepker, M.D.
Oregon Health Sciences University	Pre- and postdoctoral training in behavioral cardiology	Robert D. Fitzgerald. Ph.D.
University of Miami (Coral Gables)	Behavioral medicine research in cardiovascular disease	Neil Schneiderman, Ph.D.
Washington University in St. Louis	Nutrition–behavioral cardiovascular disease prevention	Ruth E. Brennan, Ph.D.
New Jersey College of Medicine & Dentistry	2-Year postdoctoral training program in cardiovascular disease prevention (nutrition)	Norman Lasser, M.D.

phins on atherosclerosis, requires a unique blend of biomedical specialists and neuroscientifically-trained behavioral scientists possessing the requisite background, knowledge, and perspective to address the multidimensional issues involved in such research.

The most potentially valuable source of applicants is the large supply of students from the many behavioral science disciplines who *currently* are in predoctoral and postdoctoral programs not now directly related to biobehavioral science. The experience of department heads and their faculties, as well as NHLBI training program directors during the past five years, indicates that many talented graduate students now in training *could* be recruited to transfer into areas of interest to the Institute *if* the numbers of NHLBI predoctoral and postdoctoral traineeships in the behavioral sciences could be increased.

Informed estimates are that the discipline of psychology alone, with 3000 new doctoral students being graduated annually, quite likely has the potential to supply a pool of 300–500 highly qualified predoctoral and postdoctoral NHLBI trainees per year, given adequate funds.

Although the other behavioral science disciplines (e.g., sociology, anthropology, and economics) have fewer numbers of graduate students currently in training than does psychology, these sister disciplines quite likely could attract *over a period of 3–5 years* possibly as many as 50–100 predoctoral and postdoctoral trainees.[1] Although the National Academy of Sciences (4) recommended that our country shift from its former emphasis on predoctoral research training to a mix of about 70 percent postdoctoral and 30 percent predoctoral traineeship support, it must be remembered that this split was an *average* across all biomedical and behavioral scientific disciplines. Clearly, the unique circumstances in one or another discipline might necessitate a different mix of numbers of pre- and postdoctoral traineeships.

4. Biobehavioral Training Options

A research student who wants to become a biobehavioral scientist now has several options. He or she may apply for an individual postdoctoral training grant or may choose to join a currently-established program at one of ten institutions (Table 1).

4.1. Individually-Funded Opportunities

Several different kinds of grants or awards are available in both public and private sectors.

1. National Research Service Award (NRSA) for Individual Postdoctoral Fellows (5)

These NIH fellowships are awarded for one, two, or three years and provide a graduated stipend (based on prior years of relevant experience) to the recipient together with some costs paid to the sponsoring institution. Applicants must have received an advanced degree such as a Ph.D., M.D., or equivalent; and the proposed training must "encompass biomedical or behavioral research training with an opportunity to carry out supervised research and offer opportunity to research health scientists, research clinicians, etc., to broaden their scientific background or to extend their potential for research in health-related areas" (5).

[1]Sociology graduates some 600 new Ph.D.s per year and Anthropology some 400 annually (personal communication from American Sociological Association, 1981). These two total 1,000 per year, or 3,000–5,000 graduates during a 3–5 year period.

Each Institute at NIH has delineated areas of interest for which funding is available. Within the NHLBI, the Division of Heart and Vascular Disease (DHVD) funds research training in basic processes, behavioral studies (e.g., risk factor modification), prevention studies, and clinical investigation.

2. NRSA for Senior Fellows (6)

The NIH fellowships provide support (up to $30,000 per year) for those with advanced degrees such as Ph.D., D.D.S., etc. who have also had at least seven subsequent years of relevant research or professional experience, and who wish to obtain further research training experience in biomedical and behavioral research. This award is given usually for one year but it may on occasion be renewed for a second year. In most other respects it shares similar characteristics with the NRSA for Individual Postdoctoral Fellows.

3. Research Career Development Award (RCDA) (7)

"The RCDA is a special NIH salary grant to enhance the research capability of individuals in the formative stages of their careers who have demonstrated the outstanding potential for contributing as independent investigators to health-related research" (p. 1). Although candidates must have a doctoral degreee and three or more years of subsequent research or professional experience, they must not be sufficiently advanced in their career as to have achieved senior faculty status or to have published extensively.

An RCDA is awarded for a continuous period of five years (to a maximum of $30,000 per year); recipients must devote all their time to research or related activities, though these may include giving or receiving research training.

4. The Merck Fellowship Award, Given by the American College of Cardiology (8).

This is a relatively new fellowship given to support advanced training in cardiology. In fiscal years 1981–82 there were 5 awards of a $15,000 stipend. To be eligible, an applicant must currently be in his or her first or second year of adult cardiology training. The American College of Cardiology is interested in those applicants who wish to do research in a unique clinically-oriented area or in a broader area such as the teaching and delivery of cardiovascular health care. Each fellowship is awarded for one year, though successful recipients may receive an additional year of funding.

5. The Clinician Scientist Award, Given by the American Heart Association (AHA) (9)

Currently there are approximately ten recipients of this award, each receiving $30,000 annually for a period of 3 years (a further 2-year extension is possible). It is awarded to young scientists (under age 35) who already have an M.D. or an M.D./Ph.D. combination. Such recipients must pursue research in a cardiovascular field including stroke and related basic science issues.

6. Grant-in-Aid (9), Given by the AHA

Between 80 and 118 grants are given each year by the AHA to talented young research scientists who have already completed a doctorate. Each recipient must demonstrate scientific excellence and must do research in the field of cardiovascular function in disease or related basic problems involving physiology, biochemistry, epidemiology, or psychology.

7. Established Investigatorship (9), Given by the AHA

This year the AHA gave 40 investigatorships—an increase over prior years. This program is well-established and has similar breadth to the Clinician Scientist Award. It has a term of 5 years and is given to young (under age 40) physicians and scientists with 3 or more years of postdoctoral experience.

5. Review of Current NRSA Institutional Research Training Programs in Cardiovascular Behavioral Medicine

The NIH funds domestic, nonprofit, private or public institutions which support research training programs for predoctoral and postdoctoral candidates. The programs "must encompass supervised biomedical or behavioral research and offer opportunity for research training leading to the research degree, or, for those who have already attained the research degree, opportunity to broaden their scientific background" (10).

Since 1976, the development of cardiovascular behavioral medicine training programs has accelerated dramatically. As noted in Table 1 and Figure 1, the institutional training programs have increased in number from one to ten over a four-year period.

1. Cornell Medical College: Training Program in Central Neural Control of the Circulation

This two-year program admits postdoctoral fellows with an M.D. and/or Ph.D; and predoctoral fellows from the Cornell University Graduate School of Medical Sciences who are already at work on a Ph.D. or M.D./Ph.D. in neurobiology and behavior.

The program provides training at the interface of neuroscience and cardiovascular medicine. Each trainee is assigned a mentor and selects a senior

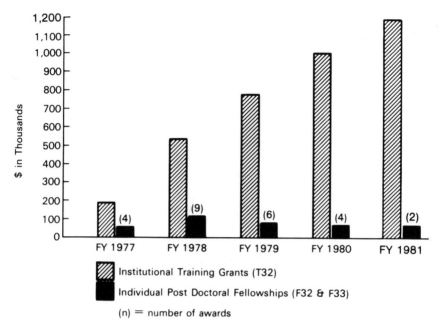

Figure 1: Biobehavioral research training in the Division of Heart and Vascular Disease funded by the National Heart, Lung, and Blood Institute.

scientist under whose direction a research project is undertaken. Each such project must involve at least two disciplines with the view of exposing trainees not only to cardiovascular physiology but also to neuroanatomy, neuropharmacology, neuropsychology, neuroendocrinology, and biochemistry.

2. Johns Hopkins University: The Cardiovascular Risk Reduction Training Program

Both predoctoral and postdoctoral fellows receive multidisciplinary training related to issues of patient, public, and professional health education; occupational health; health services research; psychophysiological aspects of cardiovascular health; program design, implementation, and evaluation; primary and secondary prevention techniques; school health; and health behavior change.

The Hopkins faculty conducts a year-long Seminar on Behavioral and Educational Aspects of Cardiovascular Risk Reduction at which trainees present their research along with that of the faculty and invited guests.

3. Stanford University: Cardiovascular Disease Prevention Training Program

This program offers multidisciplinary postdoctoral training in the broad field of cardiovascular disease prevention. Training is provided primarily through direct experience in an existing large multidisciplinary research resource—The Stanford Heart Disease Prevention Program (SHDPP). The SHDPP offers field and laboratory studies in cardiovascular epidemiology, coronary heart disease intervention trials, family aggregation studies, clinical research in cardiovascular risk factors, studies on adherence to drugs and diet, and behavioral studies in coronary heart disease and hypertension. The largest and most varied SHDPP element is a five-city study of multifactor cardiovascular risk intervention using various methods of public health education including mass media and face-to-face methods. This project offers an interdisciplinary milieu of trainees to engage in a wide variety of biostatistical, epidemiological, behavioral, and nutritional studies.

4. University of California, Berkeley: Behavioral Factors in Cardiovascular Disease Etiology

Trainees and fellows with demonstrated accomplishment in the behavioral or biomedical fields undertake two years of training leading to the M.P.H. degree. Students strong in biomedical sciences are encouraged to stress psychosociological training and vice versa. Examples of research opportunities associated with this program include sex differences in coronary heart disease; Type A behavior; risk factor identification in patients undergoing angiography; self-help efforts toward stopping smoking; and psychosocial aspects of hypertension.

5. University of Houston: Social Psychology in Prevention of Cardiovascular Disease

Program evaluation and research design skills are major elements of the Houston program. Pre- and postdoctoral trainees receive general grounding in social psychology as well as exposure to such topics as the conscious and conditioned control of cardiovascular function, patient compliance with medical regimens; advances in lipid and atherosclerosis research, cardiovascular risk reduction strategies, and hypertension control. They attend weekly seminars at Baylor College of Medicine's National Heart and Blood Vessel Research and Demonstration Center. They may become involved in research programs being conducted at Baylor concerning Type A behavior, diet modification, community hypertension control, prevention of smoking among adolescents, etc.

6. University of Minnesota: Research Training Program in Cardiovascular Health
   Behavior

Started in 1978, this program serves both predoctoral and postdoctoral trainees and emphasizes prevention, treatment, and recovery from cardiovascular disease. Seminars typically include such topics as cardiovascular physiology, epidemiology, and health behavior. Predoctoral trainees may elect course work according to their goals and field of interest in such areas as cardiovascular epidemiology, biometry, cardiovascular physiology, nutrition, psychophysiology, communication, human motivation, and social psychology.

Postdoctoral fellows are encouraged to participate in collaborative research with faculty and program consultants, and experts in education, behavior, and medicine from universities in both the United States and Europe.

7. Oregon Health Sciences University: Predoctoral and Postdoctoral Training in
   Behavioral Cardiology

This program offers training in biopsychology with specialization in behavioral cardiology. Trainees may collaborate with research cardiologists and become experienced in biofeedback techniques as well as in specialized surgical, pharmacological, and evaluation methods.

In the first year, predoctoral students study medical psychology. During their first and second years, they can choose among various medical school courses in anatomy, microbiology, physiology, pharmacology, biochemistry, etc. More advanced study of developmental biology, cardiovascular pathophysiology, and neurophysiology is offered in the later stages of their training program. Throughout the training period students engage in cardiovascular research and complete a thesis. Postdoctoral fellows must have a strong background in biopsychology and engage in research and teaching.

8. University of Miami: Behavioral Medicine Research in Cardiovascular Disease

Pre- and postdoctoral candidates receive training in the etiology, pathogenesis, treatment, and prevention of cardiovascular disease—from a biobehavioral and psychosocial standpoint. Trainees also get intensive exposure to microcomputer processing for data analysis and for use in electrophysiology.

This program, sited in the Department of Psychology, also draws upon the resources of the medical school faculty (e.g. epidemiology, pediatrics, medicine, etc.). Clinical experience is obtained at several local hospitals and clinics, including a child development center.

Current research programs span a broad spectrum of topics, from ultrastructural changes occuring in the cardiovascular system in response to short-term emotional stress to physiological reactivity in college students as a function of psychosocial characteristics (e.g. Type A behavior), behavioral challenge, and exercise. Formal coursework includes human psychophysiology, the social psychology of health behaviors, and behavioral medicine.

9. Washington University: Postdoctoral Research Training Program in Nutrition— Behavioral Cardiovascular Disease Prevention

Educators from the departments of psychology, anthropology, economics, social work, preventive medicine, cardiology, nutrition, metabolism, and pediatrics all lend their expertise to this postdoctoral program begun in 1980. Approximately 85% of trainee time is spent on a research project chosen from amongst smoking cessation studies, exercise programs for hypertensive adolescents, the role of early malnutrition on subsequent behavior, and similar topics.

10. New Jersey College of Medicine and Dentistry

This program provides postdoctoral training to nutritionists and behavioral and biomedical scientists in the behavioral aspects of cardiovascular disease prevention. Trainees are exposed to the interactive interdisciplinary milieu (epidemiology, biostatistics, internal medicine, nutrition, behavioral science) concerned with research on the prevention of coronary heart disease. This two-year program will prepare trainees for independent research and/or teaching careers in applying behavioral science principles to preventive cardiology, particularly in the area of nutrition. Opportunities for coursework and applied experience are provided through local resources (e.g., Rutgers University School of Applied and Professional Psychology), ongoing research efforts in prevention with both adults (MRFIT) and children (Know Your Body program from the American Health Foundation), as well as smoking cessation and industrial health promotion programs.

The ten training programs listed above provide a broad spectrum of research training in biomedical and behavioral science settings. Some of the programs are relatively new, while others are more established. What is evident is the rich variety of opportunities and experiences offered by the institutions involved.

6. Conclusions

The most conservative estimate of the numbers of new *research behavioral scientists* needed to both capitalize on the research knowledge recently acquired and also to begin to help meet the Surgeon General's 1990 health goals for our nation (11, 12, 13) would have to be in the neighborhood of 400–500 doctoral-level biobehaviorists. Taking the lower of these two figures, an estimated breakdown would be as follows:

	Employment site	Number of new behavioral scientists needed
(1)	U.S. Medical schools (one per school)	120
(2)	U.S. Schools of public health	50
(3)	Research positions in addition to (1) and (2); e.g. universities, research institutes, etc.	100
(4)	Government	20
(5)	Business and industry	100
		390

To acquire these 390 *new behavioral scientists* for research careers in cardiology, the current number of such scientists in training (58) would have to be increased six or sevenfold. Candidates *are* attracted to this field, encouraged by the incentive of intellectual stimulation—the incentive of wishing to be associated with an emerging dynamic area of theoretical, conceptual, and empirical development. We must be sure that every effort is made to provide sufficient training capacity and support.

It is to be hoped that other universities and institutions (both in the private and the public sectors) will establish programs similar to those described above. They can surely draw upon the wealth of experience of these initial ten, and of the NHLBI. Such healthy growth is what this nation needs if we are to be in a position by 1990 to deal adequately with the contingencies of this nation's cardiovascular health.

References

1. *Arteriosclerosis 1981: Vol 1.* Report of the Working Group of Arteriosclerosis of the National Heart, Lung and Blood Institute. NIH Publication #81-2034, July 1981.
2. *Annual Reports of the Committee on a Study of National Needs for Biomedical and Behavioral Research Personnel.* Committee Reports of the National Research Council, Washington, D.C. 1978. Title: "Personnel Needs and Training for Biomedical and Behavioral Research."
3. *The Workshop of Establishing Research Training Programs in Behavior and Health.* Ebert-Flattau, Pamela, Editor. Proceedings of the Workshop of Establishing Research Training Programs in Behavior and Health, National Research Council, Commission on Human Resources, Washington, D.C. 1981.

4. *Research Training Programs in Behavior and Health.* Proceedings for the Workshop on Establishing Research Training Programs in Behavior and Health, National Academy Press, 1981.

5. *National Research Service Awards for Individual Postdoctoral Fellows, March 1, 1980* U.S. Government Printing Office: 1980—311-201:3047.

6. *National Research Service Awards for Senior Fellows,* NIH Guide for Grants and Contracts: 8(11), August 13, 1979.

7. *Policy Brochure and Additional Instructions for Preparing an Application for a National Institute of Health Research Career Development Award.* U.S. Government Printing Office: 1981 828/8206.

8. *Merck Fellowship Award Information Brochure.* The American College of Cardiology, Heart House, 9111 Old Georgetown Road, Bethesda, Maryland 20814.

9. *Research Grants Brochure* The Director of Research Awards, The American Heart Association, 7320 Greenville Avenue, Dallas, Texas 75231.

10. *National Research Service Awards for Institutional Grants.* August 1, 1980. Announcement from the Division of Research Grants, National Institutes of Health, Bethesda, Maryland 20205.

11. *Healthy People: The Surgeon General's Report on Health Promotion and Disease Prevention, 1979.*

12. *Promoting Health—Preventing Disease: Objectives for the Nation: The Surgeon General's Report on Health Promotion and Disease Prevention, 1980.*

13. *Health United States 1980: With Prevention Profile: The Surgeon General's Report on Health Promotion and Disease Prevention, 1981.*

Part VI

Summary: Biobehavioral Perspectives in Coronary Arteriosclerosis

Chapter 13

Summary: Biobehavioral Perspectives in Coronary Arteriosclerosis

J. Alan Herd

The influence of behavioral processes on arteriosclerosis is evident in pathogenesis of the basic process, management of clinical manifestations, and prevention of arteriosclerotic cardiovascular disease. Recent progress in all these areas has come from interdisciplinary research in which the interactions between physiological and behavioral processes have been explored. Basic neurobiologic studies have indicated mechanisms whereby behavioral processes influence metabolic and hemodynamic processes. Principles of behavior analysis have revealed basic determinants of human behavior in reduction of cardiovascular disease. Development approaches toward behaviors related to cardiovascular disease have achieved some success in reducing incidence of addictive cigarette smoking in youngsters. Finally, comprehensive community programs for control of cardiovascular risk factors have been successful in reducing incidence of cardiovascular disease. Thus, the record shows that a substantial proportion of our recent progress in basic knowledge, management, and prevention of arteriosclerotic cardiovascular disease has been the result of progress in biobehavioral research.

Associations have been established indicating possible links between brain function, behavior, physiological (especially neuroendocrine) processes, and arteriosclerotic cardiovascular disease. More information, however, is needed concerning the pathophysiological links. In particular, the genetic determinants and the developmental processes influencing behavior must be determined, as well as the psychological processes whereby behavior influences arteriosclerosis.

A great challenge for the future is the application of basic principles of behavior to management of patients with arteriosclerotic cardiovascular dis-

J. Alan Herd • Sid W. Richardson Institute for Preventive Medicine, The Methodist Hospital, 6565 Fannin, MS S400, Houston, Texas 77030.

ease and to prevention of arteriosclerosis. High-risk behaviors are common even in patients with clinical manifestations of coronary heart disease and in individuals with high levels of risk factors for cardiovascular arteriosclerotic disease. Our ability to alter high-risk behaviors depends on understanding of human learning and analysis of antecedent, concurrent, and consequent events associated with risk behavior and change.

1. Etiology and Pathogenesis of Arteriosclerotic Cardiovascular Disease

1.1. Basic Process of Arteriosclerosis

Neuroendocrine Function and Arteriosclerosis

The association between neuroendocrine processes and arteriosclerosis is suggested by several clinical and experimental studies. Troxler, Sprague, Albanese, Fuchs, and Thompson (1977) found a significant correlation between elevated serial morning plasma cortisol levels and moderate-to-severe coronary arteriosclerosis. This association was noted in asymptomatic male subjects who had coronary angiography as part of their evaluation at the USAF School of Aerospace Medicine. In these men, plasma cortisol was second only to serum cholesterol as a discriminator between coronary disease and nondiseased individuals. Rheumatoid arthritis patients treated with corticosteroids are reported to have a threefold increase in arteriosclerosis (Kalbak, 1972), and corticosteroids have been reported to accelerate coronary atherosclerosis in patients with lupus erythematosus (Bulkley and Roberts, 1975). Finally, Friedman, Byers, and Rosenman (1972) have reported that plasma corticotropin (ACTH) levels were significantly higher in Type A subjects than in Type B subjects.

The influence of cortisol on development of artherosclerosis was tested experimentally in cynomolgus monkeys. Sprague, Troxler, Peterson, Schmidt, and Young (1980) administered cortisol orally to monkeys each day in doses which significantly diminished the diurnal variations of serum cholesterol without elevating daily peak cortisol concentrations. Animals fed a high cholesterol diet and given daily cortisol had a significantly greater involvement of aortic intimal surface area with atherosclerotic lesions compared to animals receiving a high cholesterol diet only. This atherogenic effect occurred independently of any effect of cortisol on serum or lipoprotein cholesterol concentrations.

A number of studies have been undertaken in attempts to identify psychophysiological and neuroendocrine mechanisms which might account for the increased coronary heart disease (CHD) rates observed among Type A persons (Review Panel on Coronary-Prone Behavior and Coronary Heart Disease, 1981). When challenged to perform a variety of behavioral tasks, Type As were observed to show greater increase in heart rate, blood pressure, and catecholamine secretion. Among those psychological characteristics that have been implicated in the tendency to display overt Type A behavior under challenge are increased need for control (Glass, 1977), increased need for self involvement (Scherwitz, Berton and Leventhal, 1978), and increased levels of hostility (Glass, 1977).

Behavior and Endocrine Function

The concept underlying the neuroendocrine correlates of behavioral processes arose with Cannon (1936) and was elaborated by Selye (1976). The original concept was one of a general physiological response including adrenal medullary, adrenal cortical, and sympathetic nervous system responses to physical and psychological stimuli. Refinements of the original concept have included identification of different physiological and behavioral responses to acute and chronic conditions and different responses to anticipated and current stimulation (Rose, 1980). Mason (1964; 1968) in reviewing the results of neuroendocrine research on the pituitary-adrenal-cortical system concluded that most studies of acute anticipated stimulations showed an increase in cortisol levels. However, physiological and behavioral responses to chronic conditions are more variable than responses to acute stimulation (Mason, 1964; Bourne, Rose, and Mason, 1967; Friedman, Mason, and Hamburg, 1963).

Mechanisms whereby behavioral processes might influence neuroendocrine activity have been demonstrated through recent studies of neuropeptides. This new family of neurochemicals includes many substances first identified as hormones secreted by the pituitary gland such as ACTH or as hormones secreted by the hypothalamus such as thyrotropic-releasing factor, the molecule by which the hypothalamus regulates through the pituitary the functions of the thyroid gland (Guillemin, Yamazaki, Justisz, and Sakiz, 1962).

Although the neurobiologic links between the behavioral processes and arteriosclerosis are still unknown, a general concept is emerging. Physiological and psychological stimuli that are aversive modify both adrenal medullary and adrenal cortical activity through secretion of neuropeptides as well as by direct neural influences. These neuropeptides also modify behavior and may be conditioned by environmental stimuli to be secreted even in the absence of painful stimuli. Adrenal medullary and adrenal cortical secretions appear to influence the development of arteriosclerosis and precipitation of complications. However, the pathophysiological mechanisms influencing arteriosclerosis under these conditions are still largely unknown.

1.2. Epidemiologic Studies of Behavioral Factors and Arteriosclerotic
     Cardiovascular Disease

The strongest demonstration of a relationship between any psychosocial factor and CHD is showing a prospective association between the Type A behavior pattern and increased risk of CHD events. Type A persons are characterized by high levels of achievement striving, speed and impatience, and aggressive, hostile behavior. In a prospective study of over 3,000 middle-aged men who were healthy at intake, the Western Collaborative Group Study found that those who were Type A experienced about twice as many clinical CHD events over an $8\frac{1}{2}$ year follow-up in comparison to their nonType A, or Type B, counterparts (Rosenman, Brand, Jenkins, Friedman, Straus, and Wurm, 1975; Rosenman, Brand, Sholtz, and Friedman, 1976). This prospective association between Type A behavior pattern and increased CHD rates has now also confirmed the Framingham study (Haynes, Feinleib, Levine, Scotch, and Kannel, 1978).

The status of Type A behavior pattern as a CHD risk factor was recently affirmed by an NHLBI-convened panel of behavioral and biomedical scientists who reviewed the available evidence concerning Type A behavior and CHD (Review Panel on Coronary-Prone Behavior and Coronary Heart Disease, 1981) and concluded that Type A behavior pattern confers increased risk of developing clinically apparent CHD, and that this increased risk " . . . is over and above that imposed by age, systolic blood pressure, serum cholesterol, and smoking, and appears to be of the same order of magnitude as the relative risk associated with any of these other risk factors."

There is evidence that stressful life events are likely to be followed by a wide variety of illnesses, including clinical CHD events. The impact of stressful life events appears to be modulated by the presence or absence of social support networks in the individual's current environment (Berkman and Syme, 1979). The relationship between social ties and mortality was assesssed using the 1965 Human Population Laboratory survey of a random sample of about 7,000 adults in California and a subsequent nine-year mortality follow-up. People who lacked social ties were more likely to die in the follow-up period than those with more extensive contacts. The age-adjusted relative risks for those most isolated when compared to those with the most social contacts were 2.3 for men and 2.8 for women. The association between social ties and mortality was independent of self-reported physical health status at the time of the initial survey, year of death, socioeconomic status, and health practices such as smoking, alcoholic beverage consumption, obesity, physical activity, and utilization of preventive health services.

Other studies of CHD rates among Japanese migrants to Hawaii and California have also documented the increased risk attributed to relatively weak social support networks. It was found that increased CHD rates among Japanese migrants living in California could not be accounted for by differences in age, diet, serum cholesterol, blood pressure, or cigarette smoking (Marmot and Syme, 1976). However, it was found in this study that those who retained traditional Japanese lifestyles and associations were far less likely to suffer clinical CHD events, leading to the inference that the high levels of social support present in traditional Japanese culture were somehow protective.

2. Diagnosis and Treatment of Arteriosclerotic Cardiovascular Disease

2.1. Cardiovascular Risk-Factor Reduction

Recently our success in treatment of patients has been increased using principles of learning in teaching patients to stop smoking cigarettes, alter eating habits, take medications as prescribed, and to follow their doctor's advice. Principles of conditioning have been used in identifying and controlling stimuli, rewards, and punishments that influence behaviors related to cardiovascular risk factors. Principles of social learning have been used in training adolescents to resist social pressures to begin smoking. Progress in treatments for obesity and cigarette smoking has been greatest when procedures have included alteration of the environment, instruction in relevant behaviors, specific rewards for desired behaviors, and attention to social influences.

Natural support systems can be effective in influencing behavior. This is pertinent to the requirements for behavior change in smoking cessation, weight control, or long-term adherence to antihypertensive medication. Social support systems facilitate the development of coping strategies that help people contain distress within tolerable limits, maintain self-esteem, preserve interpersonal relationships, meet the requirements of new situations, and prepare for the future (Hamburg and Killilea, 1979). An area for future research is the experimental construction of social support networks where natural ones are inadequate. It is plausible that social support networks, which have been prominent in human evolution for many thousands of years as a major feature of behavioral adaptation, exert their influences on health and disease largely through neuroendocrine mechanisms.

The task of interrupting a gratifying behavior pattern on which the individual has become highly dependent is very difficult. Until recently, few well-trained investigators addressed such problems. Research on smoking cessation utilizing learning principles shows that several techniques can produce 15–25 percent one-year abstinence rates (Orleans, 1980). In the United States alone, however, more than 30 million cigarette smokers have quit smoking successfully, most of them on their own with little outside aid. For adult males, there has been a decrease from rates of cigarette smoking in 1955, when 52.6 percent of all males over the age of 18 years were cigarette smokers, to rates in 1980 of approximately 35 percent (Matarazzo, 1982). Apparently, the one-year abstinence rates of 15–25 percent are obtained in treatment of cigarette smokers who have failed to quit on their own and finally seek outside aid from treatment clinics.

Interest has been rapidly increasing in utilizing the workplace for smoking cessation, but little research has so far been done in this context. The workplace appears promising for several reasons (Danaher, 1979): 1) Most adult smokers can be reached in the workplace; 2) among males who smoke, the problem is heavily concentrated among blue-collar workers who tend not to seek outside help, although surveys indicate that their desire to stop smoking is similar to that of other smokers; 3) working women smoke at a somewhat higher level than do women in the home; 4) for workers in occupations involving asbestos, cotton, uranium, and coal mining, there is special elevation of risk for smokers; and 5) the workplace provides an environment for long-term reinforcement of efforts to maintain nonsmoking behavior.

Control of Obesity

Behavioral treatment for weight control has been studied extensively. In general, treatment effects have been modest with few reports of sustained weight loss. Reports by Stuart (1967) indicated that 80 percent of a group of obese subjects lost more than 20 pounds and 30 percent lost more than 40 pounds. However, the majority of studies in weight control have been less dramatic.

Intensive behavioral treatment for weight control has been reported to be most successful. Musante and colleagues at Duke University (Musante, 1976) have provided three meals each day to obese subjects in a clinic dining room with cognitive retraining of eating habits and practical experience under supervision. Of these obese subjects who stayed in treatment for six to eleven months, 85 percent lost more than 20 pounds and 61.5 percent lost more than 40 pounds. The long-term effects of this treatment remain to be seen.

Additional studies have been performed in which effects of behavior therapy have been combined with pharmacotherapy and couples training in a 16-week behavioral weight-reduction program (Brownell and Stunkard, 1981). In these studies, patients who received fenfluramine lost significantly more weight than patients who did not receive medication, but patients in the medication group regained weight much more rapidly during the 12-month maintenance phase. The spouse conditions did not differ in weight change during treatment or follow-up. However, obese spouses lost as much weight as the patient and patients with obese spouses lost more weight than patients with nonobese spouses. It was also noted that depression increased in proportion to decreases in weight. Results of these studies indicate that initial weight loss and maintenance of that loss may require different interventions. Although pharmacotherapy increased initial weight loss, it did not improve results for long-term weight loss.

Learning new patterns of eating behavior can be facilitated by viewing such behavior in its social context. Epidemiologic research has shown that the prevalence of obesity bears a strong relationship to social class, ethnicity, and religious background. The body weight that a person maintains is in part a response to social norms and pressures (Rodin, 1978). In addition to long-term effects of social environment on eating and body weight, momentary environmental stimuli also influence eating behavior and body weight.

2.2. Clinical Management of Arteriosclerotic Cardiovascular Disease

Behavioral and sociocultural factors deserve careful attention to clinical management and rehabilitation of arteriosclerotic cardiovascular disease. Principles of behavior analysis have potentially important implications for a wide variety of treatment and rehabilitation concerns, including modification of life-styles, compliance with medical regimens, and direct treatment of angina and high blood pressure. A consideration of psychosocial characteristics which identify subgroups with greater likelihoods of survival or achievement of pain relief could help to make more rational choices of treatment for a given patient, e.g., with regard to surgical *vs.* medical management. Finally, evidence is accumulating that behavioral and psychosocial characteristics of patients are significant in determining successful rehabilitation.

2.3. Psychosocial and Behavioral Factors in Rehabilitation

With regard to rehabilitation, or secondary prevention, further potential benefits to be gained from the application of behavioral science knowledge and techniques are becoming evident. Recent research (Oldridge, Wicks, Hanley, Sutton and Jones, 1978) has identified behavioral characteristics which appear to identify reliably future noncompliers: Type A behavior patterns, depression, and cigarette smoking. In addition, there are behavioral characteristics which have been found to be related to outcomes of the rehabilitation process: Patients who were depressed prior to entry in a rehabilitation program were found on follow-up to have poor social readjustments, lower rates of return to work, and greater requirements for medical care (Stern, Pascale, and Ackerman, 1977). In addition, patients who had had a previous myocardial infarction and who were Type A, were significantly more likely on follow-up in the Western Collaborative Group Study to suffer a reinfarction than were Type B men (Jenkins, Zyzanski, and Rosenman, 1976).

This last observation led Friedman and others to undertake a prospective five-year study in which 900 Type A men who had recently had a myocardial infarction were recruited to participate in a program designed to prevent the occurrence of subsequent coronary events. Six hundred men were randomly assigned to an intensive behavior modification program designed to reduce all aspects of the Type A behavior pattern. The other 300 men were assigned to groups who received conventional education regarding the need to reduce traditional risk factors and group psychotherapy to deal with problems of anxiety and depression. Such research can sharpen the focus for rehabilitation efforts in the future.

3. Prevention of Arteriosclerotic Cardiovascular Disease

Perhaps the greatest challenge for biobehavioral science is the prevention of artherosclerosis. At present the greatest efforts have been made to influence such risk factors for cardiovascular disease as hypertension, cigarette smoking, and hypercholesterolemia. These efforts involve public education, personal counseling, group interactions, conditioning procedures, and training in techniques to resist social pressures. Results of these studies have shown that the greatest success in reduction of cardiovascular risk factors comes when social rewards are built into the natural environments in which people live.

For example, the challenge of influencing cigarette smoking has resulted in efforts to prevent children from becoming cigarette smokers rather than persuading cigarette smokers to stop smoking (Botvin, Eng, and Williams, 1980; McAlister, Perry, and Maccoby, 1979; Hurd, Johnson, Pechacek, Bast, Jacobs, and Luepker, 1980). Studies by Evans and his colleagues suggest that adolescent smokers do believe that cigarette smoking is unhealthy (Evans, 1976). However, studies by several groups of investigators indicate that peer pressure, models of smoking parents, and messages in mass media outweigh the belief of children that smoking is dangerous (Evans, Rozelle, Mittelmark, Hansen, Bane, and Havis, 1978). Thus, information alone has little effect in deterring cigarette smoking. Apparently, rather than depending primarily on communicating information to adolescents concerning the dangers of smoking, adolescents must be trained to cope with social influences on smoking behavior.

Training adolescents to resist social pressures to begin smoking has been accomplished. Studies by McGuire (1974) suggested that specific communications can be effective. Since teenagers are most likely to begin smoking during the junior high school years, Evans and his colleagues trained students in specific ways of coping with the immediate social influences to smoke. Video-tapes depicting adolescents in common social situations were presented to illustrate ways of resisting social pressure to smoke (Evans, Rozelle, Dill, Guthrie, Hanselka, Henderson, Hill, Maxwell, and Raines, 1980). Tests of smoking behavior included measures of nicotine in saliva as well as self-reports. Follow-up studies indicated substantial reductions in smoking among students in the treatment groups compared to control groups.

3.1. The Stanford Heart Disease Prevention Program

This interdisciplinary research effort has shown that decreasing risk factors associated with cardiovascular disease is possible through media-based health education (Farquhar, Maccoby, Wood, Brietrose, Haskell, Meyer, Maccoby, Alexander, Brown, McAlister, Nash, and Stern, 1977; Farquhar, 1978). The risk factors addressed in this two-year community program included cigarette smoking, obesity, high plasma cholesterol concentration, and high blood pressure. The experiment involved three communities: 1) one with a mass media education program; 2) one with a mass media program supplemented with face-to-face counseling for high-risk individuals; and 3) one control community with no special programs. During the two-year program there was a 30 percent reduction of overall cardiovascular disease risk—almost all of which was achieved in the first year of the program and sustained through the second year.

These and other findings have implications for further research. A well-planned mass media effort may be effective by itself but take longer to achieve results than when mass media communications are combined with intensive individual instruction of high-risk persons. Effects are probably better sustained if community leaders and family members are involved, thus building a social support network that can provide continuing reinforcement for health-promoting behavior. This project suggests the utility of combined efforts of biomedical research, public health, and behavioral science. New projects of this sort have recently been begun and established studies are reaching maturity.

3.2. The North Karelia Program

A broader approach to social influences on behavior also has proven effective in reducing cardiovascular risk factors. Starting in 1972, a community program for control of cardiovascular disease was begun in North Karelia, Finland (Puska, Tuomilehto, Salonen, Neittaanmaki, Maki, Virtamo, Nissinen, Koskela, and Takalo, 1979; Salonen, Puska, and Mustaniemi, 1979; Puska, Tuomilehto, Salonen, Neittaanmaki, Maki, Virtamo, Nissinen, Koskela, and

Totako, 1981). The intervention goal was reduction of cardiovascular risk factors with the expectation that morbidity and mortality from cardiovascular disease also would be reduced. Because elevated risk factor levels in most of the population are influenced by behavior, it was assumed risk factors would be linked to social factors. Consequently, a comprehensive community program was initiated using mass media, general health education measures, provision of practical services, training of local personnel, environmental changes, and installation of communication and information systems. All these activities were supported enthusiastically by governmental officials and by the community.

The effect of the program on risk factors, its costs, and its social impact were evaluated by examining independent representative population samples in North Karelia and another matched reference area. More than 10,000 subjects were studied at the outset in 1972 and after five years in 1977. Using a multiple logistic function for computing effects of cigarette smoking, concentrations of serum cholesterol, and levels of arterial blood pressure, a significant reduction in risk of 17.4 percent was observed among men and of 11.5 percent among women in North Karelia compared to the reference area. At the same time, there was reduction in incidence rate of acute myocardial infarction among men of 16.7 percent and of 10.2 percent among women. Incidence of cerebrovascular accidents declined 12.7 percent among men and 35.5 percent among women. Mortality from all causes also declined. The costs of the program were small and matched well with the evident benefits for the community.

Although average reductions in cigarette smoking, serum cholesterol values, and levels of blood pressure were small, the impact of the program on cardiovascular disease was substantial. Results of this community-based intervention trial indicate the effectiveness of social networks in diffusing information, initiating behaviors, and maintaining reductions in cigarette smoking, serum cholesterol values, and levels of blood pressure (McAlister, Puska, Salonen, Tuomilehto, and Koskela, 1982).

3.3. The Oslo Study

Recent reports from the Oslo study (Hjermann, 1980; Hjermann, Velve-Byre, Holme, and Leren, 1981) indicate the success that can be achieved in reducing coronary risk factors using counselling and group instruction in stopping smoking and changing food habits. This five-year, randomized controlled clinical trial has admitted 1,232 high-risk men, aged 40–49 years with serum cholesterol values 290–380 mg/dl, of whom 80 percent initially were daily cigarette smokers. At the outset of the program 604 men were allocated to the intervention group and 628 men to the control group.

During the five years of the study, the counseling and group instruction induced not only a considerable change in blood lipids and smoking patterns, but also in other variables related to atherosclerotic diseases. Mean serum cholesterol concentrations were approximately 13 percent lower in the intervention group than in the control group during the trial. Mean fasting serum triglycerides fell by 20 percent in the intervention group compared with the controls. In addition, the tobacco consumption decreased by 45 percent more in the intervention group than in the control group. However, only 25 percent of the smokers in the intervention group completely stopped smoking compared with 17 percent in the control group. At the end of the trial, the incidence of myocardial infarction and sudden death was 47% lower in the intervention group than in the controls. Statistical analyses showed that the reduction in incidence in the intervention group was correlated with the reduction in total serum cholesterol concentrations and to a lesser extent with smoking reduction. Thus, the reduction of a combination of two important risk factors demonstrated the effects of intervention in reducing incidence of myocardial infarction and sudden death (Hjermann et al., 1981).

4. Prospects for the Future in Biobehavioral Research

Objectives for behavioral medicine research in arteriosclerosis should include (1) determining the pathophysiological links of behavior to arteriosclerosis; (2) increasing effectiveness of behavioral interventions; and (3) preventing development of health-damaging behaviors in adolescents.

Investigations of pathophysiological links must be made if there is to be progress in relating behavior to arteriosclerosis. In the biomedical sphere, the influence of sympathetic nervous system function, neuroendocrines, and cardiovascular factors in arteriosclerosis must be determined. In the behavioral sphere, the physiological and biochemical concomitants of psychological and social processes must be determined. Neurobiological, physiological, and behavioral studies have been carried out in several species under many different conditions. However, all processes seldom have been studied in the same species under the same conditions. Coordinated correlative studies should be developed in relevant animal species including human subjects. Furthermore, the physiological responses to behavioral processes must be determined using techniques that do not disturb either the physiological or the behavioral process under study. Fortunately, noninvasive techniques have been developed for studies of cardiovascular and metabolic processes in unanesthetized subjects. Also, technology is available for analyzing chemical substances such as hormones and metabolites in small quantities of tissue and fluid. At present, there is great promise for research on the neuroendocrine influences in arteriosclerosis using in vivo and in vitro techniques to determine influences of metabolic, endocrine,

and neural substances on platelet aggregation, vascular smooth muscle proliferation, lipoprotein receptor density, and other processes known to influence
arteriosclerosis.

Techniques for increasing the effectiveness of behavioral interventions
should be developed. In particular, the maintenance of desired health-related
behaviors over long periods of time should be improved. Efforts should be made
to determine the motivators most powerful in maintaining desired behaviors.
These include tangible rewards as well as intangible social and personal benefits. In addition, techniques are required to identify behaviors, quantify those
behaviors, influence them, evaluate their outcome, and determine the direct
and incidental consequences of those behaviors on physiological, psychological,
and social processes. Techniques for eliciting conditioned responses, cognitive
responses, emotional responses, and techniques for studying coping processes
need to be developed. In particular, efforts must be made to develop more
sophisticated techniques for eliciting psychological and behavioral processes
that can be graded in intensity, repeatable on retesting with habituation, and
comparable in effects when used by different investigators. The examples provided by exercise physiologists in developing graded exercise tolerance tests
should be emulated by psychologists who could enhance our capabilities for
studying psychological and behavioral processes under controlled conditions.

Finally, preventing development of health-damaging behaviors in adolescents is an important challenge for behavioral scientists. Training adolescents
to resist social pressures to begin smoking has been undertaken. Studies suggest
that adolescent smokers do believe cigarette smoking is unhealthy. However,
studies indicate that peer pressure, models of smoking parents, and messages
in mass media outweigh the belief of children that smoking is dangerous. Thus,
information alone has little effect in deterring cigarette smoking. Studies also
must be conducted concerning the development of other health-related behaviors such as eating habits, exercise, and development of behavior patterns similar to Type A patterns seen in adults. The influence of family, education, peers,
media, and other social forces should be studied for their effects on acquisition
of health-promoting and health-damaging behaviors in young children and
adolescents.

This biobehavioral approach to arteriosclerosis depends upon (1) objective
measures of human and animal behavior; (2) quantitative measures of physiological and biobehavioral variables associated with behavioral phenomena;
and (3) behavioral techniques for initiating and maintaining behaviors over
long periods of time. We are on the threshold of great advances using biobehavioral approaches in arteriosclerosis research. Attention will be directed
toward genetic, developmental, and pathophysiological factors as well as social,
psychological, and environmental factors influencing behavior. As we broaden
our scope we will increase our contribution to the prevention and control of
arteriosclerotic cardiovascular disease.

References

Berkman, L. F., and Syme, S. L. Social networks, host resistance, and mortality: A nine-year follow-up study of Alameda County residents. *American Journal of Epidemiology*, 1979, *109*, 186–204.

Bernstein, D. A., and McAlister, A. The modification of smoking behavior: Progress and problems. *Addictive Behavior*, 1976, *1*, 89–102.

Botvin, G., Eng, A., and Williams, C. L. Preventing the onset of cigarette smoking through life-skills training. *Preventive Medicine*, 1980, *9*, 135–143.

Bourne, P. G., Rose, R. M., and Mason, J. W. Urinary 17-OHCS levels. Data on seven helicopter ambulance medics in combat. *Archives of General Psychiatry*, 1967, *17*, 104–110.

Brownell, K. D., and Stunkard, A. J. Couples training, pharmacotherapy, and behavior therapy in the treatment of obesity. *Archives of General Psychiatry*, 1981, *38*, 1224–1229.

Bulkley, B. H., and Roberts, W. C. The heart in systemic lupus erythematosis and the changes induced in it by corticosteroid therapy. *American Journal of Medicine*, 1975, *58*, 243–264.

Cannon, W. B. The role of emotion in disease. *Annals of Internal Medicine*, 1963, *9*, 1453–1465.

Danaher, B. G. Smoking cessation programs in occupational settings: "State of the art" report. In: *National conference on health promotion programs in occupational settings*. U. S. Department of Commerce, HRP-0030860, 1979.

Eisinger, R. A. Psychosocial predictors of smoking recidivism. *Journal of Health and Social Behavior*, 1971, *12*, 355–362.

Evans, R. I. Smoking in children: Developing a social psychological strategy of deterrence. *Journal of Preventive Medicine*, 1976, *5*, 122–127, 1976.

Evans, R. I. Rozelle, R. M., Dill, C. A., Guthrie, T. J., Hanselka, L. L., Henderson, A. H., Hill, P. C., Maxwell, S. E., and Raines, B. E. The Houston project: Focus on target-based filmed interventions. In: *Symposium on Deterrents of Smoking in Adolescents: Evaluation of Four Social Psychological Strategies*. Montreal, Que., Canada: American Psychological Association, 1980.

Evans, R. I., Rozelle, R. M., Mittelmark, M. B., Hansen, W. B., Bane, A. L., and Havis, J. Deterring the onset of smoking in children: Knowledge of immediate physiological effects and coping with peer pressure, media pressure, and parent modeling. *Journal of Applied Social Psychology*, 1978, *8*, 126–135.

Evans, R. I., Rozelle, R. M., Maxwell, S. E., Raines, B. E., Dill, C. A., Gurthrie, T. J., Henderson, A. H., and Hill, P. C. Social modeling films to deter smoking in adolescents: Results of a three-year field investigation. *Journal of Applied Psychology*, 1981, *66*, 399–414.

Farquhar, J. W. The community-based model of life style intervention trials. *American Journal of Epidemiology*, 1978, *108*, 103–111.

Farquhar, J. W., Maccoby, N., Wood, P. D., Brietrose, H., Haskell, W. L., Meyer, A. J., Maccoby, N., Alexander, J. K., Brown, B. W., McAlister, A. L., Nash, J. D., and Stern, M. P. Community education for cardiovascular health. *Lancet*, 1977, *1*, 1192–1195.

Fishbein, M. *Consumer beliefs and behavior with respect to cigarette smoking: A critical analysis of the public literature*. Report prepared for the staff of the Federal Trade Commission, 1967.

Friedman, M., Byers, S. O., and Rosenman, R. H. Plasma ACTH and cortisol concentration of coronary-prone subjects. *Proceedings of the Society for Experimental Biology and Medicine*, 1972, *140*, 681–684.

Friedman, S. B., Mason, J. W., and Hamburg, D. A. Urinary 17-hydroxycorticosteroid levels in parents of children with neoplastic disease: A study of chronic psychological stress. *Psychosomatic Medicine*, 1963, *25*, 364–376.

Glass, D. C. *Behavior patterns, stress, and coronary disease*. Hillsdale, N. J.: Lawrence Erlbaum Associates. 1977.

Guillemin, R., Yamazaki, E., Justisz, M., and Sakis, E. Presence dans un extraite de tissus hypothalamiques d'une substance stimulant la secretion d l'hormone hypophysaire thyreo-trope (TSH). Premiere purification par filtration sur gel Sephadex. *Compte Rendus Heb-domadaires des Seances de L. Academie des Sciences. D: Sciences Naturelles (Paris),* 1962, *255,* 1018–1020.

Hamburg, B., and Killilea, M. Relation of social support, stress, illness, and use of health services. In: *Healthy people: the Surgeon General's Report on Health Promotion and Disease Prevention, Background Papers.* DHEW publication No. (PHS) 79-55071A. U. S. Government Printing Office, 1979, pp. 253–276.

Haynes, S. G., Feinleib, M., Levine, S., Scotch, N., and Kannel, W. B. The relationship of psychosocial factors to coronary heart disease in the Framingham Study, II. Prevalence of coronary heart disease. *American Journal of Epidemiology,* 1978, *107,* 384–402.

Hjermann, I. Smoking and diet intervention in healthy coronary high risk men. Methods and five-year follow-up of risk factors in a randomized trial. The Oslo Study. *Journal of Oslo City Hospitals,* 1980, *30,* 3–17.

Hjermann, I., Velve-Byre, K., Holme, I., and Leren, P. Effect of diet and smoking intervention on the incidence of coronary heart disease. Report from the Oslo Study Group of a randomized trial in healthy men. *Lancet,* 1981, *2,* 1303–1310.

Hurd, P. D., Johnson, C. A., Pechacek, T., Bast, L. P., Jacobs, D. R., and Luepker, R. V. Prevention of smoking in seventh grade students. *Journal of Behavioral Medicine,* 1980, *3,* 15–28.

Jenkins, C. D., Zyzanski, S. J., and Rosenman, R. H. Risk of new myocardial infarction in middle-aged men with manifest coronary heart disease. *Circulation,* 1976, *53,* 342–347.

Kalbak, K. Incidence of arteriosclerosis in patients with rheumatoid arthritis receiving long-term corticosteroid therapy. *Annals of Rheumatic Diseases,* 1972, *31,* 196–200.

Mahoney, Michael J. Behavior modification in the treatment of obesity. *Psychiatric Clinics of North America,* 1978, *1,* 651–660.

Marmot, M. B., and Syme, S. L. Acculturation and coronary heart disease in Japanese-Americans. *American Journal of Epidemiology,* 1976, *104,* 225–247.

Mason, J. W. A review of psychoendocrine research on the pituitary–adrenal cortical system. *Psychomatic Medicine,* 1968, *30,* 576–607.

Mason, J. W. Psychoendocrine approaches in stress research. In *Symposium on medical aspects of stress in the military climate.* Walter Reed Army Institute of Research, Washington, D. C., 1964

Matarazzo, J. Behavioral health's challenge to academic, scientific, and professional psychology. *American Psychologist,* 1982, *37,* 1–14.

Mausner, B. An ecological view of cigarette smoking. *Journal of Abnormal Psychology,* 1973, *81,* 115–126.

McAlister, A., Puska, P., Salonen, J. T., Tuomilehto, J., and Koskela, K. Theory and action for health promotion. Illustrations from the North Karelia Project. *American Journal of Public Health,* 1982, *72,* 43–50.

McAlister, A. L., Perry, C., and Maccoby, N. Adolescent smoking: Onset and prevention. *Pediatrics,* 1979, *63,* 650–658.

McGuire, W. J. Communication-persuasion models for drug education: Experimental findings. In M. Goodstadt (Ed.), *Research on methods and programs of drug education.* Toronto, Ontario, Canada: Addiction Research Foundation. 1974.

Musante, G. J. The dietary rehabilitation clinic: Evaluative report of a behavioral dietary treatment of obesity. *Behavioral Therapy,* 1976, *7,* 198–204.

Nash, J. D., and Farquhar, J. W. Community approaches to dietary modification and obesity. *Psychiatric Clinics of North America,* 1978, *1,* 713–724.

Office of Cancer Communications. *The smoking digest: Progress report on a nation kicking the habit.* National Cancer Institute, Bethesda, Maryland, 1977.

Oldridge, M. B., Wicks, J. R., Hanley, C., Sutton, J. R., and Jones, N. L. Noncompliance in an exercise rehabilitation program for men who have suffered myocardial infarction. *Canadian Medical Association Journal,* 1978, *118,* 361–364.

Orleans, C. S. Summary of presentation—Quitting smoking: Promising approaches and critical issues. In *Institute of Medicine, Health, and Behavior: A research agenda interim report no. 1; smoking and behavior,* 1980.

Puska, P., Tuomilehto, J., Salonen, J. T. Neittaanmaki, L., Maki, J., Virtamo, J., Nissinen, A., Koskela, K., and Totako, T. The North Karelia Project: Evaluation of a comprehensive community program for control of cardiovascular disease in 1972–77 in North Karelia, Finland. Geneva: WHD Monograph Series, 1981.

Puska, P. Tuomilehto, J., Salonen, J. T., Neittaanmaki, L., Maki, J., Virtamo, J., Nissinen, A., Koskela, K., and Takalo, T. Changes in coronary risk factors during comprehensive five-year community program to control cardiovascular diseases (North Karelia project). *British Medical Journal,* 1979, *2,* 1173–1178.

Rodin, J. Environmental factors in obesity. *Psychiatric Clinics of North American,* 1978, *3,* 581–592.

Rose, R. M. Endocrine responses to stressful psychological events. *Psychiatric Clinics of North America,* 1980, *3,* 251–276.

Rosenman, R. H., Brand, R. J., Jenkins, C. D., Friedman, M., Straus, R., and Wurm, M. Coronary heart disease in the Western Collaborative Group Study. *Journal of the American Medical Association,* 1975, *233,* 872–877.

Rosenman, R. H., Brand, R. J., Scholtz, R. E., and Friedman, M. Multivariate prediction of coronary heart disease during 8.5 year follow-up in the Western Collaborative Group Study. *American Journal of Cardiology,* 1976, *37,* 903–910.

Salonen, J. T., Puska, P., and Mustaniemi, H. Changes in morbidity and mortality during comprehensive community program to control cardiovascular diseases during 1972–77 in North Karelia. *British Medical Journal,* 1979, *2,* 1178–1183.

Scherwitz, L., Berton, K., and Leventhal, H. Type A behavior, self-involvement, and cardiovascular response. *Psychosomatic Medicine,* 1978, *40,* 593–609.

Selye, H. *The Stress of Life.* New York: McGraw-Hill, 1976.

Sprague, E. A., Troxler, R. G., Peterson, D. F., Schmidt, R. E., and Young, J. T. Effect of cortisol on the development of atherosclerosis in cynomolgus monkeys. In S. S. Kalter (Ed.), *The Use of Nonhuman Primates in Cardiovascular Diseases.* Austin: University of Texas Press, 1980.

Stern, M. J., Pascale, L., and Ackerman, A. Life adjustment postmyocardial infarction. *Archives of Internal Medicine,* 1977, *137,* 1680–1685.

Stuart, R. B. Behavioral Control of overeating. *Behavioral Research and Therapy,* 1967, *5,* 357–365.

The Review Panel on Coronary-Prone Behavior and Coronary Heart Disease. Coronary-Prone behavior and coronary heart disease: A critical review. *Circulation,* 1977, *26,* 151–162.

Troxler, R. G., Sprague, E. A., Albanese, R. A., Fuchs, R., and Thompson, A. J. The association of elevated plasma cortisol and early artherosclerosis as demonstrated by angiography. *Atherosclerosis,* 1977, *26,* 151–162.

U. S. Public Health Service. *Teenage smoking: National patterns of cigarette smoking, ages 12 through 18, in 1972 and 1974.* DHEW publication No. (NIH) 76-931. U. S. Department of Health, Education & Welfare, Public Health Service, National Institutes of Health, 1976.

Index